Retrieval of Medicinal Chemical Information

W. Jeffrey Howe, EDITOR
The Upjohn Company

Margaret M. Milne, EDITOR
Smith, Kline, and French

Ann F. Pennell, EDITOR
ICI Americas, Inc.

Based on a symposium cosponsored
by the Divisions of Computers in
Chemistry and Chemical Information
at the 175th Meeting of the
American Chemical Society, Anaheim,
California, March 13–17, 1978.

ACS SYMPOSIUM SERIES **84**

AMERICAN CHEMICAL SOCIETY
WASHINGTON, D. C.   1978

Library of Congress CIP Data

Symposium on Retrieval of Medicinal Chemical Infor-
mation, Anaheim, Calif., 1978.
Retrieval of medicinal chemical information.
(ACS symposium series; 84 ISSN 0097–6156)

Based on a symposium cosponsored by the Divisions
of Computers in Chemistry and Chemical Information
at the 175th meeting of the American Chemical Society, California, March 1978.
Includes bibliographies and index.

1. Information storage and retrieval systems—Chemistry, Pharmaceutical—Congresses. 2. Chemistry, Pharmaceutical—Data processing—Congresses.
I. Howe, William Jeffrey, 1946– . II. Milne, Margaret M., 1946– . III. Pennell, Ann F., 1946– .
IV. American Chemical Society. Division of Computers
in Chemistry. V. American Chemical Society. Division of Chemical Information. VI. Title. VII. Series:
American Chemical Society. ACS symposium series; 84.

RS421.S93 1978   615′.19′02854   78-21611
ISBN 0–8412–0465–9   ACSMC 8 84 1-231 1978

ACS Symposium Series

Robert F. Gould, *Editor*

FOREWORD

The ACS SYMPOSIUM SERIES was founded in 1974 to provide a medium for publishing symposia quickly in book form. The format of the Series parallels that of the continuing ADVANCES IN CHEMISTRY SERIES except that in order to save time the papers are not typeset but are reproduced as they are submitted by the authors in camera-ready form. Papers are reviewed under the supervision of the Editors with the assistance of the Series Advisory Board and are selected to maintain the integrity of the symposia; however, verbatim reproductions of previously published papers are not accepted. Both reviews and reports of research are acceptable since symposia may embrace both types of presentation.

CONTENTS

PREFACE

The symposium on retrieval of medicinal chemical information was organized to examine current developments in the storage, retrieval, and manipulation of the variety of types of data that are associated with medicinal chemistry in the pharmaceutical industry, government agencies, and related organizations. This volume contains expanded versions of the papers presented at the symposium as well as several additional invited papers.

To insure adequate coverage of what has become a broad and increasingly important field, the speakers were selected to approach the topic from a number of different viewpoints, discussing such subjects as the manipulation of biological data, chemical substructure searching, computer graphical display of retrieved data, the integration of biological search results with chemical information, the utilization of retrieval systems in the research function, mathematical analyses of chemical-structure data bases, and so on. While some of the chapters deal with commercially available information systems, most focus on the capabilities of systems that were developed within individual organizations. Authors were encouraged to include not only what has been done in the area, but what is now being planned for implementation in the near future to meet the growing information needs of medicinal chemical research.

Several papers have been included that were not presented at the symposium. These invited submissions extend the treatment of the subject beyond the limitations of a one-day symposium. However, the breadth of the field of medicinal chemical information has made it impossible to offer complete coverage in a volume of this size. For example, the storage and retrieval of clinical test data is one important area which could not be dealt with here. Early in the planning stages of the symposium it was recognized that some form of overview of medicinal chemical information would be a valuable addition to a proceedings volume. To that end we have written an introductory chapter, based on the contents of the symposium presentations and on discussions with symposium participants. It is hoped that this overview will do three things: (a) provide a fairly complete statement of the current status and directions of progress of the field as a whole, (b) illustrate the interrelationship of the various categories of medicinal chemical information, including those important areas which could not be dealt with in the symposium, and (c) provide a conceptual framework for viewing the material discussed in the papers which follow.

One other point deserves mention. The ACS Books Department has instituted a policy of peer review for papers included in the ACS Symposium Series. Although this adds time and effort to the publication process, it can only result in an improvement in the quality of the papers, and will benefit both readers and authors. We fully support this move.

We wish to thank the officers of the Divisions of Chemical Information and Computers in Chemistry for their assistance, especially Gabrielle Revesz, Mary Reslock, and Ed Olson.

The Upjohn Company
Kalamazoo, MI 49001

W. JEFFREY HOWE

Smith, Kline, and French
Philadelphia, PA 19101

MARGARET M. MILNE

ICI Americas, Incorporated
Wilmington, DE 19897

ANN F. PENNELL

August 22, 1978

Retrieval of Medicinal Chemical Information—an Overview

MARGARET M. MILNE—Smith, Kline, and French, Philadelphia, PA 19101

ANN F. PENNELL—ICI Americas, Inc., Wilmington, DE 19897

W. JEFFREY HOWE—The Upjohn Company, Kalamazoo, MI 49001

The following paper was written by the organizers of the ACS Symposium on Retrieval of Medicinal Chemical Information and is based in part on the symposium presentations, on discussions with symposium participants, and on the authors' own involvement in the field of pharmaceutical research and development.

Traditionally, the term "medicinal chemistry" has connoted an area of synthetic organic chemistry which deals with the preparation of molecules likely to have some desired physiological response. Associated with each synthesized molecule is a collection of *in vivo* or *in vitro* test results used to ascertain the actual nature and extent of the bioactivity (if any). Evolving in parallel with this view, medicinal chemical information systems have commonly been based on a data file organized by compound and have contained such items as chemical structure, identification number, source, and sometimes physical properties. The biological test results were also organized by compound, but either because of their volume or for administrative reasons the results were usually separate from the structural data file. Over the past 15 years considerable effort has been invested in computerizing these files, in developing efficient, powerful, and rapid mechanisms for selective retrieval, and in integrating the searching of structural data with that of biological data without actually combining the individual files.

More recently, however, it has been recognized that the traditional view of medicinal chemistry is actually one element of a much larger set of functions in the total drug development process. These functions are interrelated and at times interdependent, and the drug development process can be made markedly more effective by facilitating the necessary interactions. What this implies to developers and users of medicinal chemical information systems is a need for access to a considerably more diverse set of drug-related information types and for additional capabilities in retrieving, correlating, and displaying these data. This, in fact, is the direction of current progress in the field of medici-

nal chemical information. To examine this progress in more detail, the individual functions that comprise the total drug development process and the corresponding information needs will
now be considered.

The Drug Development Process

Figure 1 illustrates schematically a hypothetical drug development program. The diagram is approximate since the sequence
of functions may vary somewhat between different organizations
and since not all of the conceivable interactions among functions
are indicated. In addition, some of the functions may overlap
(e.g., pathology/toxicology studies may continue during clinical
trials) and some may be done in segments interleaved among the
other functions (e.g., applications for Food and Drug Administration (FDA) approval may occur at various stages). However, the
diagram does include all of the major functions and their approximate relationships.

Development of a drug begins with selection of a prototype
or lead compound or compound series whose activity is to be opti-
mized. Commonly an organization such as a pharmaceutical company
has certain major areas of interest (e.g., antimicrobials, con-
traceptives, etc.) within which new lead structure types are
sought. For each such area of interest, a set of biological
tests or screens is designed specifically to test for the desired
activity. The lead compounds are normally found either through
random testing in these screens of diverse chemical types or
through ideas triggered by public or in-house literature.

The selected lead is then developed by a cyclic process in
which analogs are synthesized and bioassayed, results are analyzed, and new analogs postulated to have even greater activity
are proposed for synthesis. This is the traditional realm of
medicinal chemistry and the area in which most of the work on
medicinal chemical information retrieval has been done. Novel
compounds synthesized in the analog evaluation may be patented at
any point, and for particularly promising compounds, additional
analogs may be prepared for patent protection.

Compounds that show particularly good activity in the primary biological screening tests are submitted to pathology, toxicology, and pharmacology studies to further define their suitability as drugs. For those that show promise for clinical use appropriate formulations are developed. When a compound satisfactory in all of these areas has been found, a request (Investigation New Drug application, or IND) to test the drug in humans is
submitted to FDA summarizing all existing data on the compound.
More or less concurrently, process development studies are undertaken to optimize the manufacturing process and for patent protection. Marketing implications may also be considered in
further detail at this point.

When the IND application is approved, clinical trials can be

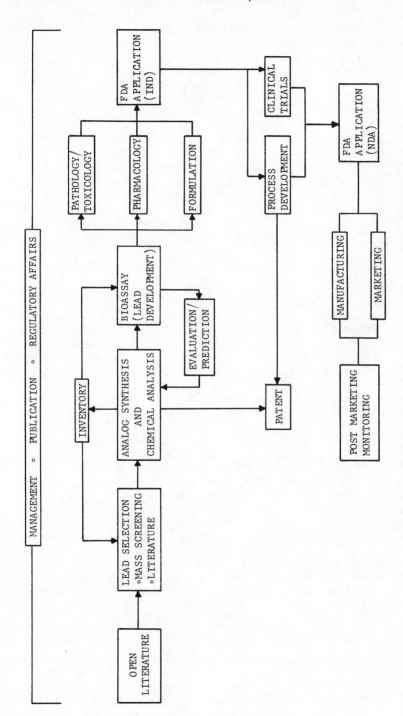

Figure 1. Major functions in a typical drug-development program

initiated. Concurrently, manufacturing facilities and a marketing program are established. If the clinical testing is successful, the results are submitted to the FDA along with a New Drug
Application (NDA) requesting approval of the drug for use in the
general population. Finally, once the product is on the market,
its use is monitored to detect any additional indications of use
or adverse effects that may not have been evident during original
testing.

In addition to these directed functions, three types of
functions exist which pervade essentially all of the other functions of Figure 1: (1) the management function in which decisions are made to determine which projects or compounds will
progress, which will be delayed or terminated; (2) the publication function in which results are provided to the internal and
open literature; and (3) the function of responding to government
regulations relating to chemical research and manufacturing.

Information Needs Related to the Drug Development Process

Development of a safe and useful drug, which is the ultimate
goal of medicinal chemistry, is an extremely complex and costly
process. The purpose of medicinal chemical information retrieval
is to support this process, to permit rapid and accurate identification of clinically useful compounds with minimal risk, cost,
or delay. The types of information required to provide this support are extensive and diverse. The exact types of data needed
will be discussed in more detail later, but some generalizations
can be made about the way the data must be used.

The basic data functions required are information storage,
retrieval, analysis, and reporting. The organization of the data
into various files must be balanced for maximum efficiency. Although each of the functions of Figure 1 has its own primary interest information, many situations require the combined use of
data from different areas. For example, the management, publishing, and regulatory affairs functions require access to nearly
all of the data types at one point or another. Thus, while the
data must be segmented to allow efficient access by its prime
users, interfaces must be provided to satisfy cross-functional
needs as well.

The quality of the data is a vital characteristic. "Quality"
here encompasses not only accuracy, but comprehensiveness (inclusion of old and new data) and suitability (providing the exact
type of data needed rather than some nebulous function or quali-
tative estimate thereof). Comprehensiveness is particularly
important where chemicals with human bioactivity are being prepared.

Procedures for using the system should be as attractive and
as simple as possible to encourage direct use by specialists within each function who may not also be information specialists.
Interactive operation, simple commands, flexible outputs that are

familiar to the user (such as standard text rather than computer codes, standard structural diagrams rather than linear notations) all help to get the information directly into the hands of the individuals best equipped to use it. Although the provision of interfaces suitable for non-computer specialists adds considerably to system development and operational costs, these costs should be far outweighed by the resulting increase in effectiveness of the total program. (Of course, some functions are still unavoidably complex and will require intermediation by an information specialist for the forseeable future.)

Organization and Current Status of Medicinal Chemical Information

The major data types required for medicinal chemistry as part of a total drug development program are indicated in Figure 2. In this conceptualization the data types are organized as a network around the chemical compound, which is identified by a unique number (generally an internal registry number). Four categories of data are defined: chemical data and biological data, which characterize the compound itself, management/distribution data which characterize the commercial aspects of the compound, and secondary or bibliographic data which in essence are pointers to chemical, biological, or management/distribution data in the open (or occasionally, internal) literature.

(a) Chemical Data. Virtually all organizations that support a major drug development effort have a computerized file of the chemical structures considered in the program. Files on the order of 100,000 - 400,000 structures are not uncommon. The structures are represented as either connection tables, line notations, or attribute codes. At present the files are used for several purposes, the most common of which are (a) duplicate checking to determine if a compound has already been tested, (b) substructure searching to selectively retrieve compound classes, and (c) display of two dimensional structural diagrams.

Of the chemical data categories in Figure 2 the molecular structure representation has received by far the greatest emphasis in current information systems. The other data types (analytical, physicochemical, process development, synthesis/reactivity) have been dealt with individually, but are only beginning to be incorporated with the structure files to form integrated systems.

(b) Biological Data. Virtually all major medicinal chemical information systems have automated files which contain the primary testing results. Because of the importance of correlating and coordinating biological data with structural data in the analog development cycle, automated links between the structure data and primary testing data have been provided in some systems. However, the sheer volume of the biological test results (and in some cases, administrative considerations) has contributed to the slow

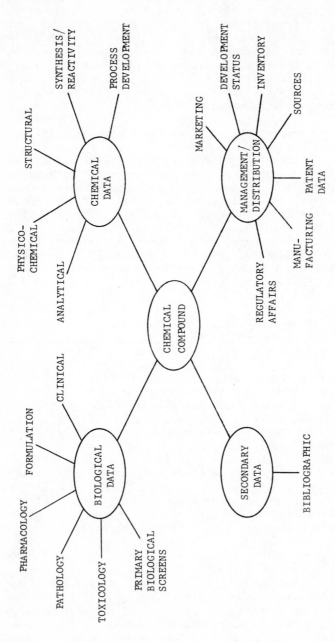

Figure 2. The medicinal chemical information network. The types of data required in a total drug-development process can be viewed as organized around the compounds tested, with four associated categories of data: chemical, biological, management/distribution, and secondary.

progress in this area. In many systems the interface between the
two data files is either unwieldly or nonexistent.

As was the case with chemical data, the remaining forms of
biological information (toxicology, pathology, pharmacology,
clinical, formulation) have all undergone a certain degree of
automation, but the extent to which these data are integrated
with chemical structures and primary screening data varies considerably. As noted above, a major constraint appears to be the
volume of data that must be encompassed by these biological systems. Interfacing them with each other and with chemical information for purposes of retrieval can place a heavy burden on even
very large computer systems. As a result, users must commonly
rely on manual methods for coordinating the data from different
areas.

(c) Management/Distribution. The management/distribution
portion of the medicinal chemical information network contains
categories of data which are extremely important to the operation
of an effective drug development program. The differences between existing systems are greatest in this area, in terms of the
data types that are handled and the capabilities and interfaces
to the rest of the network that are provided. Some of these data
types (e.g., inventory and compound sources) are generally automated and well-integrated with the chemical and biological data.
Patent information on in-house compounds may be automated but is
not commonly linked directly to the chemical and biological data.
Manufacturing is usually a separate information category, not
linked to the rest of the system. Except in special cases, such
as manufacturing problems related to the chemistry of a particular process, this separation is reasonable at present.

Marketing is another area which currently appears to be di-
vorced from the rest of the network. The importance of market
analysis (both before and after approval of a product) as an integral part of the drug development effort is becoming increasingly apparent. Such analysis is vital during lead selection to
determine what products are needed, during process development to
guarantee reasonable manufacturing costs (where "reasonable" is
related to the drug's market), and during the initial stages of
public use to respond to any effects that may not have been evident during testing.

One of the most active areas in the handling of medicinal
chemical information is concerned with the requirements of
government regulatory agencies. This includes information required not only for purposes of drug approval but also to comply
with regulations concerning the environmental effects of the manufacturing process and the health and safety of individuals ex-
posed to chemicals in any phase of the drug development process.
In many organizations major efforts are now being undertaken to
incorporate health, safety, and environmental data into the total
chemical information network.

(d) Secondary Data. Almost all drug development organizations utilize one or more of the major online information services for retrospective and current awareness literature review. In addition, in-house information groups are often responsible for reviewing the current literature on specific problems. Most categories of data that are currently retrievable through the online services are text oriented (such as abstracts, keywords, and patent data), however increasingly more specialized data types are being offered, such as physical constants, reactions, biological data (e.g., LD50), and substructure information. Some integration of these files exists, though at present the integration is primarily within a single information assembling organization. Nevertheless, there appears to be a growing recognition of the need for interfacing and compatibility even among competing services.

Government supported files are becoming an important source of public information. These files are generated to help industry respond to government regulations as well as to assist technical researchers. Data bases containing structures of pertinent compounds (e.g., regulated drugs, carcinogens, toxic compounds) are rapidly being built, and sometimes incorporate additional useful chemical information such as crystallographic and spectral data. An interesting side effect to the development of these files has been the encouragement of compatibility among private systems. This has resulted because a number of private organizations are seeking uniform methods of accessing in-house and public files and have standardized on the format of the public systems.

Future Trends in Medicinal Chemical Information

While the following chapters in this book present a comprehensive view of the current capabilities of medicinal chemical information systems, they also provide insights into the directions of progress of the field as a whole.

The foremost trend is toward integration: the pulling together of discrete in-house systems and the creation of automated interfaces to public and government systems along the lines of the information network in Figure 2. Much of the stimulus for integration comes from the cross-disciplinary nature of the technical information needs (such as the need for simultaneous access to chemical and biological data in the analog development process). At the same time, as the information systems become more encompassing and more responsive to the technical needs, they are enabling better-informed and more coordinated management decisions at higher levels. Such support of the management function is expected gradually to become more formal, offering management specific reporting and inquiry capabilities with simultaneous access to all types of data on demand.

Another extremely important trend is toward greater end-user orientation. This area was somewhat neglected while system development efforts focused on techniques for data storage and retrieval. The emphasis on end-users is evident in the increased application of computer graphics to the user-computer interface and in the growing number of interactive systems. Greater flexibility in adapting to specialized needs is apparent not only in the variety of data types and data manipulations that are possible, but also in the organization and reorganization of outputs to permit data to be examined from many standpoints. The retrieved data is more end-user oriented in order to be interpretable without cumbersome lookup tables or other artifacts of computerization. Overall, the apparent objective is to make information systems a working tool of the individuals best equipped to use them, rather than a reference utility accessible only through the information specialist.

Now that automated <u>retrieval</u> is well established, efforts are focusing on automated <u>analysis</u> of the retrieved data. Elaborate statistical and heuristic analysis procedures and sophisticated functions such as quantum mechanics and conformational analysis are being interfaced directly to large files. Interest in computer-aided synthetic analysis, computer-aided structure elucidation, and computer-aided process development remains high, but as these complex applications are still in the active development stages they have not yet undergone full-scale integration into existing information networks.

<u>Summary</u>

This chapter has outlined the nature of the drug development process and has described medicinal chemical information retrieval in terms of a network of drug-related information categories. All of the systems described in this volume can be viewed in terms of this organization. Some of the systems, notably those of the major pharmaceutical companies and some of the larger government programs, encompass a large fraction of the drug development functions and data types. Other systems may include only a single function and relatively few data types, yet even these smaller (or less integrated) systems feed into and are part of the total medicinal chemical information network.

This introductory chapter has provided only an overview of the field and a framework for viewing the specific capabilities that exist. The details and exact mechanisms are provided in the papers that follow and in the references cited therein.

RECEIVED August 29, 1978.

2

The PRODBIB Data Base: Retrieval of Product Information from the Published Literature

BARBARA C. FREEDMAN

Product Information Section, Technical Information Department, Burroughs Wellcome Co., 3030 Cornwallis Road, Research Triangle Park, NC 27709

The Library at Burroughs Wellcome Co. has traditionally been charged with the responsibility for locating, holding, and retrieving published scientific and medical literature (hereafter referred to as "product papers") about Company products. The references collected have been used in two ways: (1) to produce bibliographies to accompany Investigational New Drug applications, New Drug Applications, and Annual Reports to the Food and Drug Administration; and (2) to support the Research, Development, Medical, and Marketing Divisions in their ongoing work with Company products.

Until 1972 this responsibility was discharged using entirely manual methods. Three files existed:

    (1) a card file, organized by product and, within product, by author;

    (2) copies of bibliographies prepared from these cards since 1966; and

    (3) reprints of articles published since 1970.

About 25,000 references had been collected and the files were growing at the rate of approximately 5000 per year. Searching these manual files became increasingly difficult, as did the production of annual bibliographies. In 1972 we selected INQUIRE (Infodata Systems Inc., Falls Church, Virginia) as an appropriate software package to handle a data base of our own design, into which we could enter and retrieve bibliographic information, subject indexing, and data extracted from the product papers. We call the data base PRODBIB, for "product bibliography." Our selection of INQUIRE was based on the successful experience of other pharmaceutical firms (1).

Hardware. From November 1972 to November 1975 we ran our INQUIRE data bases at Triangle Universities Computation Center, a large, university-owned computation center. In November 1975 we moved INQUIRE operations to our own Computer Services Division, where we have an IBM 370/148, and run INQUIRE under OS/VS1, in batch mode.

0-8412-0465-9/78/47-084-010$05.00

INQUIRE facilities. We are currently using Version 9.1, with
blocked files and the multi-data base option.

Organization of the PRODBIB file. An item in the PRODBIB file
consists of the bibliographic data and indexing terms for one
product paper. If two or more products are mentioned in one
paper, however, we create one item for each product. Identical
bibliographic information is entered for each, with indexing
terms appropriate for each product.

Items are structured according to the fields definition
table (Figure 1). The only required field in the data base is
REPORT, a report code and accession number assigned to each
paper. For papers which mention more than one product, a DOCPT
(a single alphabetic character) is assigned. Consequently, REPORT
uniquely identifies the product paper; REPORT and DOCPT uniquely
identify a PRODBIB item.

The fields PATIENT through TYPE give a variety of bibliographic information, a quick categorization of each paper, and
the number of patients involved in each study. The DRUG field
contains the National Drug Code number, B.W. Co. compound
number, and B.W. Co. tradename for the product being indexed. We
use the tradename in the DRUG field only and generic names in
the remaining fields. In this way we can distinguish in searching
between the product as the subject of an item and the product
used together with or compared to another product.

The remaining fields are used for subject indexing. For
almost all of these we use terms selected from MALIMET, the
thesaurus of the Excerpta Medica Foundation (2). MALIMET is a
broad and deep list of terms, with fairly good control and a
certain amount of structure. (It is not, however, hierarchical.)
We have used about 8000 unique terms from MALIMET. Figure 2
shows the content of a typical PRODBIB item.

Indexing policy. It is our policy to index only that information
in a paper relevant to our product. We do not index the whole
content of a paper, as MEDLARS would, but rather the narrower
range of information about our product. The primary advantage
of this approach is that we can eliminate many false drops
that occur when we search the "global" data bases.

Keywords are posted against field names in such a way as to
create a keyword-in-context situation, *i.e.*, a keyword can later
be searched in terms of the field in which it was entered. For
example, the keyword RASH can be used as an indication for one
product, an adverse reaction to another product, and a contra-
indication for a third product. In searching, the keyword can be
used alone or as a field value condition of a particular field.

DISPLAY FIELDS.

FIELD NAME	KEY	TYPE	STORED LENGTH	STRUC	RPTS	- PRINT - FORM	LEN	NOTES
DATABASE 'PRODBIB '								
REPORT	PFX	CHR	12	BASE	SCALAR	NB	12	
SOURCE	PFX	CHR	4	SUBF	SCALAR	NB	4	POS 1 TO 4(REPORT)
DOCYR	PFX	CHR	2	SUBF	SCALAR	NB	2	POS 6 TO 7(REPORT)
DOCNO		CHR	4	SUBF	SCALAR	NB	4	POS 9 TO 12(REPORT)
DOCPT		CHR	1		SCALAR	NB	1	
PATIENTS		INT	4		SCALAR	I	4	
PUBYR	PFX	CHR	2		SCALAR	NB	2	
INPUT		INT	6		SCALAR	I	6	
LONDON		CHR	6		SCALAR	NB	6	
INDEXER	PFX	CHR	3		SCALAR	NB	3	
DRUG		CHR	38	BASE	V 5	NB	38	
DRUGNO		CHR	9	SUBF	V 5	NB	9	POS 1 TO 9(DRUG)
DRUGLBLK		CHR	3	SUBF	V 5	NB	3	POS 1 TO 3(DRUG)
DRUGPROD		CHR	4	SUBF	V 5	NB	4	POS 4 TO 7(DRUG)
DRUGPKG		CHR	2	SUBF	V 5	NB	2	POS 8 TO 9(DRUG)
COMPOUND	PFX	CHR	9	SUBF	V 5	NB	9	POS 10 TO 18(DRUG)
CMPONO		CHR	4	SUBF	V 5	NB	4	POS 10 TO 13(DRUG)
CMPOSRC		CHR	1	SUBF	V 5	NB	1	POS 14 TO 14(DRUG)
CMPYR		CHR	2	SUBF	V 5	NB	2	POS 15 TO 16(DRUG)
CMPBATCH		CHR	2	SUBF	V 5	NB	2	POS 17 TO 18(DRUG)
PRODNAME	SMP	CHR	20	SUBF	V 5	NB	20	POS 19 TO 38(DRUG)
AUTHOR	PFX	CHR	24		V 10	NB	24	
TITLE		CHR V	280		SCALAR	B	280	
CITATION		CHR V	168		SCALAR	B	168	
LANGUAGE	PFX	CHR	15		SCALAR	NB	15	
LOCATION		CHR	60		SCALAR	NB	60	
CATEGORY	PFX	CHR	12		V 5	NB	12	
TYPE	PFX	CHR	9		V 2	NB	9	
INDICATN	SMP	CHR	40		V 10	B	40	
EFFECT	SMP	CHR	40		V 5	B	40	
CONCUM	SMP	CHR	40		V 5	B	40	
COMPARE	SMP	CHR	40		V 5	B	40	
EXPSUBJ		CHR	45	BASE	V 5	NB	45	
EXPTYP1	SMP	CHR	20	SUBF	V 5	NB	20	POS 1 TO 20(EXPSUBJ)
EXPTYP2	SMP	CHR	20	SUBF	V 5	NB	20	POS 21 TO 40(EXPSUBJ)
EXPNO		CHR	4	SUBF	V 5	NB	4	POS 41 TO 44(EXPSUBJ)
EXPSEX		CHR	1	SUBF	V 5	NB	1	POS 45 TO 45(EXPSUBJ)
INVITRO		CHR	50	BASE	V 5	NB	50	
PREPN	SMP	CHR	30	SUBF	V 5	NB	30	POS 1 TO 30(INVITRO)
PREPSRC	SMP	CHR	20	SUBF	V 5	NB	20	POS 31 TO 50(INVITRO)
FORMULA		CHR	62	BASE	V 5	NB	62	
FORM	PFX	CHR	12	SUBF	V 5	NB	12	POS 1 TO 12(FORMULA)
RATIO		CHR	10	SUBF	V 5	NB	10	POS 13 TO 22(FORMULA)
DRUG1		CHR	10	SUBF	V 5	NB	10	POS 23 TO 32(FORMULA)
CONC1		CHR	10	SUBF	V 5	NB	10	POS 33 TO 42(FORMULA)
DRUG2		CHR	10	SUBF	V 5	NB	10	POS 43 TO 52(FORMULA)
CONC2		CHR	10	SUBF	V 5	NB	10	POS 53 TO 62(FORMULA)
DOSAGE		CHR	44	BASE	V 5	NB	44	
DOSE		CHR	40	SUBF	V 5	NB	40	POS 1 TO 40(DOSAGE)
ROUTE	PFX	CHR	4	SUBF	V 5	NB	4	POS 41 TO 44(DOSAGE)
ADVERSE		CHR	43	BASE	V 10	NB	43	
ADVREAC	SMP	CHR	40	SUBF	V 10	NB	40	POS 1 TO 40(ADVERSE)
ADVNUM		INT	3	SUBF	V 10	I	3	POS 41 TO 43(ADVERSE)
CONTRIND	SMP	CHR	40		V 5	B	40	
SENSORG	SMP	CHR	40		V 20	B	40	
RESISORG	SMP	CHR	40		V 20	B	40	
SUBJECTS	SMP	CHR	40		V 20	B	40	
EFFICACY		CHR V	252		SCALAR	B	252	
NOTES		CHR V	630		SCALAR	B	630	
KEYS		BLT	40			S,	40	
ITEMNO		BLT	8		SCALAR	NB	8	
$FIXED		CHR	109		SCALAR	NB	109	
ALLKEYS		BLT	40			S,	40	

Figure 1.   Fields definition table for the PRODBIB data base

```
$FIXED     ILBD/77/1399A002877010178         MGP
DRUG       0810852                 SEPTRA
                                   TRIMETHOPRIM
                                   SULFAMETHOXAZOLE
AUTHOR     STAMEY TA
           CONDY M
           MIHARA G
TITLE      PROPHYLACTIC EFFICACY OF NITROFURANTOIN MACROCRYSTALS AND
           TRIMETHOPRIM-SULFAMETHOXAZOLE IN URINARY INFECTIONS.
           BIOLOGIC EFFECTS ON THE VAGINAL AND RECTAL FLORA
CITATION   N ENGL J MED 296(14): 780-783 APR 7, 1977
CATEGORY   CLINICAL
           PHARMACOLOGY
TYPE
INDICATN   URINARY TRACT INFECTION PROPHYLAXIS
           RECURRENT URINARY TRACT INFECTION
EFFECT
CONCOM     CONTRACEPTIVE AGENT
           INTRAUTERINE CONTRACEPTIVE DEVICE
COMPARE    NITROFURANTOIN MACROCRYSTAL
           TRIMETHOPRIM
EXPSUBJ    HUMAN                   ADULT                    0028F
INVITRO
FORMULA    TABLET
DOSAGE     40 MG TMP + 200 MG SMX/DAY 6 MONTHS        ORAL
ADVERSE
CONTRIND
SENSORG
RESISORG
SUBJECTS   INTESTINE FLORA
           FECES MICROFLORA
           VAGINA BACTERIAL FLORA
           CONTROLLED CLINICAL TRIAL
           HYSTERECTOMY
           DRUG RESISTANCE
           ESCHERICHIA COLI
           PROTEUS MIRABILIS
           KLEBSIELLA
           PSEUDOMONAS
           ENTEROCOCCUS
           LOW DOSE
           LONG TERM THERAPY
           R FACTOR
EFFICACY
NOTES
```

Figure 2.   Content of a sample PRODBIB item

```
AUTHOR=HISS D                               1
AUTHOR=HITCHCOCK CR                         2
AUTHOR=HITCHCOCK M                          1
AUTHOR=HITCHINGS CH                         1
AUTHOR=HITCHINGS G                          2
AUTHOR=HITCHINGS GH                       153
AUTHOR=HITCHINGS GM                         1
AUTHOR=HITT BA                              1
AUTHOR=HITZENBERGER G                      18
AUTHOR=HITZENBERGER H                       1
AUTHOR=HITZIG WH                            7
```

Figure 3.   A portion of the PRODBIB author index, part of the
system-generated keyword list (with frequency counts)

```
DISPLAY REPORT=TLBD/76/1291 REPORT=TLBD/76/134).

KEYWORD DISPLAY FOR DATABASE PRODBIB
REPORT=TLBD/76/1291                         1
REPORT=TLBD/76/1292                         1
REPORT=TLBD/76/1293                         1
REPORT=TLBD/76/1294                         2
REPORT=TLBD/76/1295                         1
REPORT=TLBD/76/1296                         1
REPORT=TLBD/76/1297                         1
REPORT=TLBD/76/1298                         1
REPORT=TLBD/76/1299                         1
REPORT=TLBD/76/1300                         2
```

Figure 4.   A portion of the PRODBIB report code index, part
of the keyword list (with frequency counts)

```
DISPLAY LANGUAGE=HEBREW LANGUAGE=POLISH.

KEYWORD DISPLAY FOR DATABASE PRODBIB
LANGUAGE=HEBREW                             4
LANGUAGE=FUNGARIAN                         42
LANGUAGE=ITALIAN                          377
LANGUAGE=JAPANESE                          39
LANGUAGE=KOREAN                            3
LANGUAGE=NORWEGIAN                         18
LANGUAGE=POLISH                           66
```

Figure 5.   Language designations on the keyword list (with
frequency counts)

```
DISPLAY CARBAMAZEPINE CARCINOMA.

KEYWORD DISPLAY FOR DATABASE PRODBIB
CARBAMAZEPINE                                    7
CARBAMAZEPINE DERIVATIVE                         1
CARBAMYLPHOSPHATE SYNTHETASE                     1
CARBANILIC ACID DERIVATIVE                       3
CARBAZOLEQUINONE                                 1
CARBAZYLQUINONE                                  1
CARBENICILLIN                                   48
CARBENICILLIN DISODIUM SALT                    223
CARBENICILLIN INDANYL SODIUM                     3
CARBENOXALONE                                    3
```

*Figure 6. MALIMET subject indexing terms displayed with
frequency counts*

```
FIND SEPTRA AND CATEGORY=CLINICAL AND 'RECURRENT URINARY TRACT INFECTION',
    &BIBLIO1,
    HEADER 'SEPTRA AND RECURRENT URINARY TRACT INFECTION'.



        SEPTRA AND RECURRENT URINARY TRACT INFECTION


KUNIN CM                        TLBD/75/2958
URINARY TRACT INFECTIONS. FLOW CHARTS (ALGORITHMS) FOR DETECTION AND TREATMENT
JAMA 233(5): 458-462 AUG 4, 1975
                MENTION

LEVY SB                         TLBD/77/1694
FECAL FLORA IN RECURRENT URINARY TRACT INFECTION
N ENGL J MED 296(14): 813-814 APR 7, 1977


LINES D                         TLBD/77/1194
CHILDHOOD URINARY CANDIDIASIS SUCCESSFULLY TREATED WITH 5-FLUOROCYTOSINE
AUST PEDIATR J 12(1): 49-52, 1976


MAGANTO E                       TLBD/75/3462
PEREZ COUTINO A
TREATMENT OF REFRACTORY URINARY INFECTIONS WITH TRIMETHOPRIM-SULFAMETHOXAZOLE
ARCH ESP UROL 24(4): 369-378, 1971
        SPANISH
```

*Figure 7. PRODBIB search using keywords (search specification followed by search
results)*

```
FIND  ZYLOPRIM  AND  CATEGORY=CLINICAL  AND  INDICATN  IS  GOUT,
       &BIBLIO1,
       HEADER  'ZYLOPRIM  TREATMENT  OF  GOUT'.




                        ZYLOPRIM  TREATMENT  OF  GOUT


       GREENE  ML                          TLBD/74/2439
       GLUECK  CJ
       FUJIMOTO  WY
       SEEGMILLER  JE
       BENIGN  SYMMETRIC  LIPOMATOSIS  (LAJNOIS-BENSAUDE  ADENOLIPOMATOSIS)
       WITH  GOUT  AND  HYPERLIPOPROTEINEMIA
       AM  J  MED  48(2):  239-246  FEB  1970


       GREENE  ML                          TLBD/74/1345
       CLINICAL  FEATURES  OF  PATIENTS  WITH  THE"PARTIAL"  DEFICIENCY  OF
       THE  X-LINKED  URICACIDURIA  ENZYME
       ARCH  INTERN  MED  130(2):  193-198  AUG  1972


       GREILING  H                         TLBD/74/3371
       PRESENT  PROBLEMS  OF  GOUT.  CLINICAL  BIOCHEMISTRY  OF  GOUT
       THERAPIEWOCHE  22(2):  77-84,  1972  TRANSLATION
              GERMAN

       GUNTHER  R                          TLBD/76/3194
       KNAPP  E
       CLINICAL  FINDINGS  AND  THERAPY  OF  GOUT  WITH  SPECIAL  REFERENCE  TO  THE
       METABOLIC  EFFECTS  OF  SULFINPYRAZONE  (ANTURAN)  AND  ALLOPURINOL
       (ZYLORIC)
       WIEN  KLIN  WOCHENSCHR  81:  817-820  NOV  7,  1969
              GERMAN
```

Figure 8.   PRODBIB search using field value conditions (search specification followed by search results)

CLINICAL USES OF ALKERAN

INDICATION	ITEMS
OSTEOSARCOMA	25
OVARY ADENOCARCINOMA	2
OVARY CANCER	36
OVARY CANCER METASTASIS	1
OVARY CARCINOMA	38
OVARY CYSTADENOCARCINOMA	2
OVARY TERATOMA	1
OVARY TUMOR	2
PANCREAS ADENOCARCINOMA	1
PANCREAS CANCER	1

CLINICAL USES OF ALKERAN

MULTIPLE MYELOMA
    LATOS DL   TLBD/73/0369
    VALENTINE AM
    TREATMENT OF HYPERCALCEMIA WITH FUROSEMIDE AND CORTICOSTEROIDS
    W VIRGINIA MED J 69(3): 52-54 MARCH 1973

MULTIPLE MYELOMA
    LAW IP   TLBD/76/0609
    PLOVNICK HS
    BEDDOW DG
    MULTIPLE MYELOMA, SIDEROBLASTIC ANEMIA AND ACUTE LEUKEMIA
    N ENGL J MED 294(3): 164 JAN 15, 1976
                            LEITER

MULTIPLE MYELOMA
    LAW MIP   TLBD/76/1065
    FAMILIAL OCCURRENCE OF MULTIPLE MYELOMA
    SOUTH MED J 69(1): 46-48 JAN 1976

MULTIPLE MYELOMA
    LE CHEVALLIER PL   TLBD/72/0226
    TREATMENT OF MYELOMA.
    SEM HOP PARIS 47: 249-251 JAN 20,1971.
        FRENCH

Figure 9.   PRODBIB search listing refrences wtih a table of contents

ADVERSE REACTIONS TO SEPTRA

FOLIC ACID DEFICIENCY
    BUSHBY SRM   TLBD/74/1036
    TRIMETHOPRIM AND SULPHONAMIDES: LABORATORY STUDIES
    S AFR MED J 44(SUPPL): 3-10 AUG 15, 1970

FOLIC ACID DEFICIENCY
    CLARK F   TLBD/76/4108
    DRUGS AND VITAMIN DEFICIENCY
    ADVERSE DRUG REACT BULL 57: 196-199, 1976

FOLIC ACID DEFICIENCY
    COLMAN N   TLBD/76/4197
    HERBERT V
    COTRIMOXAZOLE AND FOLATE METABOLISM
    LANCET 2(7992): 967 OCT 30, 1976
                                    LETTER

FOLIC ACID DEFICIENCY
    DAVIS RE   TLBD/73/0753
    JACKSON JM
    TRIMETHOPRIM/SULPHAMETHOXAZOLE AND FOLATE METABOLISM
    PATHOLOGY 5(1): 23-29 JAN 1973

*Figure 10.   PRODBIB search listing adverse reactions and bibliographic
reference*

TREATMENT OF POLYCYTHEMIA VERA WITH B.W. CO. PRODUCTS

LEUKERAN
    WESTIN J   TLBD/77/2527
    CHROMOSOME ABNORMALITIES AFTER CHLORAMBUCIL THERAPY OF POLYCYTHAEMIA VERA
    SCAND J HAEMATOL 17(3): 197-204 SEP 1976

LEUKERAN
    WALLERSTEIN RO   TLBD/77/3174
    POLYCYTHEMIA VERA-WHO NEEDS THERAPY?
    CONSULTANT 17(4): 148-156 APR 1977

MYLERAN
    KAHN SB   TLBD/73/0368
    BRODSKY I
    THERAPY OF MYELOPROLIFERATIVE DISORDERS
    CANCER CHEMOTHER II ED BY I BRODSKY AND SB KAHN NEW YORK LONDON GRUNE
    AND STRATTON 1972 PP 347-359
                            REVIEW

MYLERAN
    WASSERMAN LR   TLBD/73/0273
    THE MANAGEMENT OF POLYTHAEMIA VERA
    BR J HAEMATOL 21(4): 371-376 OCT 1971
                            MENTION
                            REVIEW

Figure 11.   PRODBIB search listing various products used to treat one disease

Keyword list. The keyword list which is generated as a result
of our specifications in the fields definition table contains an
author index and a report code index (Figures 3 and 4).   In
addition, LANGUAGE, TYPE, CATEGORY, COMPOUND, and ROUTE are
prefixed keywords (Figure 5). All other fields with MALIMET
terms are keyed (Figure 6).

Data entry. We enter bibliographic data and indexing terms onto
preprinted forms. The data is then keypunched, edited from disk
using the IBM CMS editor, put on tape, and the INQUIRE updates
done from tape.

Typical search products. A majority of the searches done on the
PRODBIB file combine a drug name and one or more keywords
(Figure 7).   Less frequently, a drug name is combined with a
field value condition (Figure 8).   In addition to these
typical searches, we can produce listings arranged alphabetically
by indication or adverse reaction. These can be either lists of
the bibliographic references themselves or lists of the keywords
with a count of the number of items in which they have been
used (Figure 9).   Occasionally we produce lists of adverse
reactions occurring during treatment of particular diseases
(Figure 10). Another type of search results in listings of
references by product for the treatment of certain diseases
(Figure 11).

Utilization of the data base. We run about 50 queries per month
to answer questions received in our reference activities. We
find INQUIRE to be an extremely flexible tool for the building
of a bibliographic data base, and have been very pleased by
its ability to respond to our needs.

Literature cited.

1 Bennett, R.E. and S.J. Frycki – Internal processing of
   external reference services. J. Chem. Doc. 11(2): 76–83,
   May 1971.

2 Blanken, R.R. and B.T. Stern – Planning and design of on-line
   systems for the ultimate user of biomedical information.
   Inf. Proc. Man. 11(8–12): 207–227, 1975.

RECEIVED August 29, 1978.

3

Automation of Test-Data Transmission from Producer to Computer Master File

HELEN F. GINSBERG, DAVID J. JAMES, and CHRIS R. MONDELLO

Abbott Laboratories, North Chicago, IL 60064

Input to the Abbott Automated Biological and Chemical Data System (ABCD) (1,2) a proprietary data base operational for over a decade, was not efficient by current data handling standards. Until recently test results on internal compounds were transcribed onto appropriate code sheets, keypunched and verified, then sent to the corporate computer center for entry into the master data base. Not only were these multiple operations time consuming, but with each recording of the data there was another chance for human error.

The potential for human error can be decreased by eliminating, wherever appropriate, keypunching of large volumes of data per experiment or transcription of results from computer statistical calculations by having the experimenter enter original observations directly onto a computer terminal. All further processing of the raw data would be controlled by computer programs. Humans would intervene only when required for visual evaluation of the information or approval of outputted results. This revision in data handling procedures is currently underway.

Agricultural Research

Manual Data Entry. The agricultural research program was one area in which the recording, keypunching and processing of the test results was becoming very costly. Additionally, as illustrated in Figure 1, there was considerable shipping of paper copy from one physical location to another.

After the data were recorded on the code sheet, a small portion of which is shown in Figure 2, activity ratings were added and the sheets copied for security. Mail service between Abbott's Research Farm and our North Chicago facility was slow, sometimes resulting in a delay of several days before the test results were received for keypunching. Only after the verified data cards were read into the R&D DEC-System 10 file was further processing rapid.

0-8412-0465-9/78/47-084-020$05.00

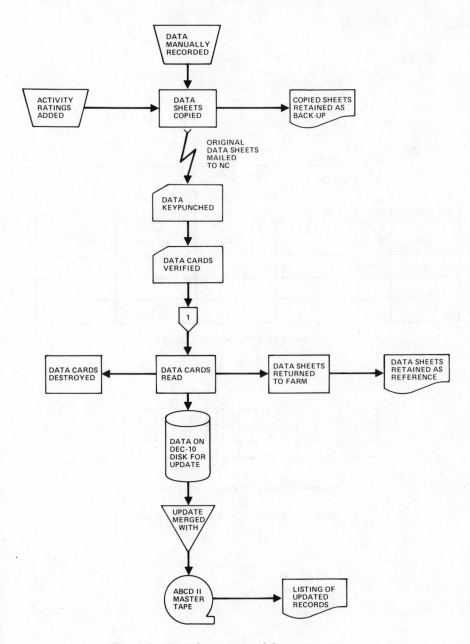

Figure 1.   Data flow in manual data entry system

A-NUMBER	SEQ. NO.	CARD 1 20—27 VELVET LEAF	28—35 SMART WEED	36—43 PIG WEED	44—51 CURLEY DOCK
1—6	74—76				

PORTION OF HERBICIDE CODE SHEET

10C4H
10C4H7Y
10G5E/3B

WRITTEN

1	0	C		4	H		
1	0	C		4	H		*
1	0	G		5	E	/	B

PUNCHED

RESPONSE FOR ONE PLANT SPECIES

Figure 2.   Detail of herbicide code sheet

An entire page of typed instructions for the keypunch operators accompanied the completed forms which had been designed originally for ease of field recording of data in herbicide and growth regulator tests. At that time the data were not being entered into a computer system. Figure 2 illustrates the detail that could be recorded for one or more of the plant species studied. Inclusive card columns for the specific fields were preprinted on the code forms. These card columns were not always in sequence, as can be seen in the figure. Additionally, for active compounds there might be more observations recorded on the sheet than could be accommodated by the eight columns allowed for each plant. An asterisk in the last column of the field, shown on the second line of the "Response for one Plant Species" would denote "Additional Information Available on the Original Record". The observer would record the response codes without specific spacing; the keypunch operator was responsible for the correct positioning of the information on the cards. Obviously keypunching would be slow. In 1975 charges for keypunching and verification for two of the tests recorded on these forms averaged $1.05 per compound.

Automated Data Entry. In mid-1975 an effort to eliminate or reduce existing manual data recording systems in the ongoing agricultural research program was begun. Although the effort was primarily directed at reducing hand coding and transcription of chemical herbicide and plant growth regulation screening results, additional applications in field plot research were also studied.

System Design Criteria. The criteria used to evaluate and design the automated data collection system were:

1. Hardware Portability - Agricultural research data is generated in a variety of locations. Screening systems may operate in the high humidity and temperature of a greenhouse as well as in the clean, climate-controlled laboratory. Field plot research may be located miles from the nearest building, and data collection must be done under all kinds of weather conditions. To operate under these conditions, the collection hardware must be self-powered, fairly tolerant of its environment and of a size and weight that allows easy carrying. Remote transmission of the collected data to a computer mainframe over normal telephone lines also is required.

2. Potential Benefits - A substantial number of existing manual data systems should be replaced. There should be saving of people hours and improved accuracy.

3. User Acceptance - Transition to a new system should be accomplished with minimal disruption in work flow. An automated system that closely resembles existing manual recording schemes will almost always gain quick user acceptance. Failure to obtain this approval will destroy any chance of successful system implementation.

4. GLP Guidelines - The system should conform to Good Laboratory Practices rules for data collection and handling (3).

5. Compatability With Existing Computer Systems - The collected data should be acceptable to existing databases. Minimal or no additional equipment or modification should be required to facilitate the transmission of data to the computer mainframe.

6. Flexibility - The system should be adaptable to the recording of various types of data. Alphanumeric capability is required. Simple methods for data identification, formatting and editing are necessary. Recording of any type of data from a single numeric entry to annotations and researchers' comments should be acceptable. Varying the kind of data collected should not require any change in the basic collection-transmission-mainframe acceptance system.

7. System Support - Reprogramming of the system should be limited to the secondary reformatting steps.

System Configuration. A two level software organization is used to attain maximum flexibility while not disturbing the basic data collection system.

1. Primary Software - Three linked system Macros serve as the basic procedures of the data collection system. The functions of these programs are:

  a. Answering the phone line, logging the job into the system, and logging the job off the system after sensing the end of file condition.

  b. Acceptance of data after assigning a unique sequential filename to the transmission and storing of the file on disk at a known location.

  c. Interpretation of the file editing and formatting characters as final preparation for their reformatting by higher level language programs.

2. Secondary Software - Several reformatting programs, written in COBOL and FORTRAN, are called by the user according to the type of data being processed. The flow of these programs allows the user to edit-in additional data such as activity ratings, corrections and comments. The use of higher level languages allows these reformatting programs to be written by programmers who are not familiar with the overall system. The only programming requirements are knowledge of the data format as it emerges from the primary software and of the final format needed by the user.

3. Hardware - The portable data entry terminal chosen for the system was an MSI Data Corporation Source 2002 (4). This battery powered device has a typewriter style keyboard that can generate the digits 0-9, letters A-Z, special characters -, +, =, *, comma, up-arrow, and three additional user requested characters. The unit records data on a digital grade cassette tape that has a capacity of 30,000 characters per side. The standard data recording code is ASCII, and it is written in a self clocking mode. Automatic detection of double keying is standard and options such as parity checking, field length checking, and check digits are available. An optional paper strip printer is available for use as a backup data recording device, to satisfy certain GLP requirements for timely hard copy, and as an indicator to the operator for tracking what is actually being entered into the machine. Total weight of the unit when equipped with the optional printer is 7 pounds. The transmission of data is character synchronous at 1200 baud. The transmission characteristics are compatable with AT&T 1001B Data Couplers and 202S Datasets, and voice grade lines are acceptable.

The computer mainframe used to support the data collection system is a DEC-System 10 which operates as a time-sharing service to the corporate research divisions of Abbott Laboratories. Communications between the data collection terminal and the mainframe are handled through a dedicated dial-up access port equipped with an AT&T 202S Dataset. A minimum amount of modification of the TOPS-10 operating system was required to completely implement reliable transmissions and acceptance of the data streams coming from the MSI equipment to the DEC-System 10.

System Operation. Actual operation of the system is very straightforward and closely resembles the pencil and paper method it replaces (See Figure 3). All data are recorded on the cassette and after verification sent via telephone lines to the DEC-10 computer. As the data are recorded, the operator simply enters the same codes, values, or whatever else was used in the

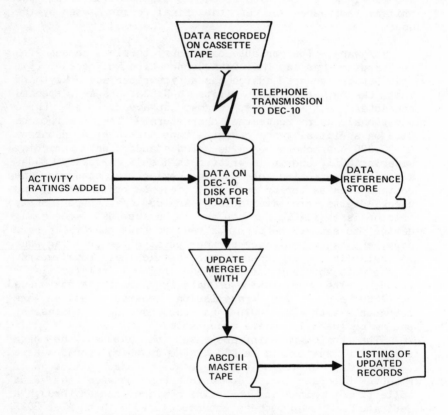

Figure 3. Data flow in automated data entry system

manual system. The only real difference is the use of special characters to take the place of printed lines to identify separate entries and lining-out to make corrections. The operator retains all the flexibility that the manual system allowed regarding comments, skipping entries, or recording complete records in any order. The only real restriction imposed upon the recording process is that each complete record must contain an exact number of entry fields whether there is a value in the field or not. This restriction is, again, no more than is found in the manual system, but it can cause some problems if the operator does not have a clear mental picture of the analogies between the automated system and the manual one.

Figure 4 illustrates the various changes a set of data undergoes in being converted from the original recorded observations until they are ready to be submitted to the master data base. The character string under TYPE MSI000.DAT is an actual representation of what appears on the MSI paper tape and the information that is transmitted to the R&D DEC-System 10 for further processing. The first ten characters preceeding the "-" sign are for transmission synchronization. The "-" sign is used to edit them out of the file. Throughout the data entry process the up-arrow ^ is used to signal "end of record." The "+" is a field delimiter; entry of "++" will force a blank field if there is no test in that run for a particular plant species which was in the sequence being studied in concurrent experiments. The "?" and "-" are used for deletion, "?" for single characters and "-" for an entire field. The next field containing "A" and numbers separated by "+" is a heading record with test date and amount of compound applied per acre (test rate). The last character string between the up-arrows includes the number of the compound being tested, second test identification and plant readings for the various species being studied.

Other information required for the complete recording of test results from the field or greenhouse are entered on the cassette in a similar manner. For example, a two-digit number is used to identify the various plant species that might be studied. Only a limited number of these would be studied at any particular time and, thus, would have to be entered with the observed results for that experiment. A "C" record is used to post the plant species for the run or runs on any day; identification numbers are recorded in the same sequence in which the results will be read and entered. This information is keyed as "C+" followed by the two-digit numbers separated by "+", then the end of record character. If there are no variations during the course of one day's readings in test dates, test rates, method of application, or test plant sequence, only one set of header records is required. If the plant sequence remains constant over a period of time, it need not be re-entered during that time interval.

```
TYPE MST000.DAT
..
<0001<0001-^A+031076+1000^24151+ 5E 9C++10E+20Y 9G++10E+ 0+ 6E+10E+ 4E+ 5E 6G+ 3G+13+GR16+ 2B 4H 6G+ 4H20??10C 2B???+ 2B+10C+ 0+ 4H

3G+ 9C+ 0+ 3H+13+^


                     Edited records
.TYPE HAVGRP.CDS


A          031076     1000

24151        5E 9C             10E        20Y 9G              10E        0          6E         10E         4E

 5E 6G   3G        13         GR16        2B 4H 6G    4H10C      2B        10C        0          4H 3G       9C          0

             3H         13


                     Intermediate formatted records used for proofing and/or documentation of test results

.TYPE SECOND.CDS

24151  15 031076 10.00  01 5E 9C      19           0310E        0420Y 9G     05           0610E        07 0

                        08 6E         0910E        10 4E        11 5E 6G     12 3G                                  13

24151  16 031076 10.00  01 2B 4H 6G   02 4H10C     03 2B        0410C        05 0         06 4H 3G     07 9C

                        08 0          09 3H                                                                         13


                     Update records for ABCD II

.TYPE OUTPUT.CDS
24151 GR1501 5E   9C     0310E        0420Y  9G     0610E        13 7603001071B
24151 GR1507 0           08 6E        0910E         10 4E           1000001072B
24151 GR1511 5E   6G     12 3G                                          001073B
24151 GR1601 2B   4H  6G 02 4H  10C   03 2B         0410C        13 7603002071B
24151 GR1605 0           06 4H  3G    07 9C         08 0            1000002072B
24151 GR1609 3H                                                         002073B
```

Figure 4. Data formats from MSI terminal to master data base

After the data are transmitted from the research farm to the R&D DEC-System 10 and the instructions transmitted with the data carried out, the edited records are transmitted back to a terminal at the farm and printed in the expanded format as shown under TYPE HAVGRP.CDS. In the tests used for this illustration a maximum of three readings could be recorded for each plant, each reading requires a length of four characters (including blanks). A "++" entered through the MSI would force a 12-position blank field in the appropriate position in the edited record. With this method of input, simple modification of entry programs allows the experimenters to record more detailed observations than were permitted with the manual data entry procedure.

The next stage in processing of the test results requires the linking of all pertinent data for these experiments prior to final reformatting for the master computer data base (ABCD). The collated records for the two herbicide tests entered through the MSI terminal are shown under TYPE SECOND.CDS. Two plant species already in the stored "C" file but not included in this experiment, "19" and "05", appear in the record. These species identifications will be deleted before final formatting for the corporate master data base. If there were, in fact, data for plant "19", a sort routine would rearrange all plant species in ascending numerical order with the appropriate information being carried along. This sort routine was included in the program series to simplify final processing of the information in the corporate data base.

All the corrected test results are stored in the DEC-10 in a compacted form until just prior to the update of the master data base. At that time these data are reformatted as shown under TYPE UPDATE.CDS. Each experiment's results are expanded into three 80 character records, control characters needed for the next series of programs added, and the records shipped to another file in the DEC-10. This latter file contains all update data that will be sent to the corporate IBM 370/158, on which ABCD resides, after a series of edits and housekeeping chores have been completed. These same herbicide test results are printed, as displayed in Figure 5, on Drug Record Sheets generated from the corporate data files.

Automation of the recording and processing of herbicide data has cut the cost of processing these test results by a factor of four (compared with the keypunching route). Additionally it has:

1. Increased the accuracy of data entry,

2. Decreased the time lag from experiment to recording in the master data base,

3. Made the experimenter totally responsible for

ABCD II DRUG RECORD PRINTED 02/18/78   COMPOUND A-24151   DAG465 - DAGR82 DPOII 794

SOURCE AMCHEM   DATE 1/62   CLINICAL DATA - NO   REQUEST NUM 1378 DEPT D912 PAGE 1
REQUESTOR D JAMES   A-24151

MOL WT 206.03
FORMULA C7 H5 CL2 N O2
NON-HAZARDOUS MATERIAL

**** PROJECT DATA ****
SCREEN ACTIVITY DOSE/CONC ROUTE SPECIES

GR15 GROUP 13 10 LBS/A

HERBICIDE TEST, PRE-EMERGENCE ADMINISTRATION:
VELVET-LEAF 5E 9C   PIGWEED 10E   CURLED DOCK 20Y 9C
OK GRASS RZ 10E   YL NUTGRASS 0   CHEATGRASS 6E
G FOXTAIL 10E   OK GRASS SD 4E   CORN 5E 6G
SOYBEAN 3G   DATE 3/76

GR16 GROUP 13 10 LBS/A

HERBICIDE TEST, POST EMERGENCE ADMINISTRATION:
TOMATO 2B 6G 4H   SNAPBEAN 4H 10C   WHEAT 2B
J GRASS SD 10C   YL NUTGRASS 0   AN MORNGLRY 4H 3G
G FOXTAIL 9C   OK GRASS RZ 0   FLD BINDWD 3H   DATE 3/76

Figure 5. Drug record sheet

validation of information before entry into the corporate data base, and

4. Given the researcher a satellite computerized personal data base.

Since the data file is initially collected in a machine readable form that has a very general format, individual researchers began to develop their own specialized data bases with this information. Within a project these scientists can search their own files, prepare their own customized reports on these data and generally access their own data from remote terminals. Inasmuch as this same information has been sent to the corporate computerized data files, they also can request searches across project lines to compare their data with other test results on compounds of interest.

Although this system was initiated by and for the agricultural research area, it is flexible enough that the basic hardware and software have been adopted by other corporate research areas for use in different types of screening systems.

Pharmacology Data

The concept of automation of information transfer from the originator to the corporate data base also has been applied to pharmacology data, especially since most of our pharmacology laboratories have terminals (hard copy and/or CRT) linked to the DEC-System 10 computer. Statisticians routinely supply programs for the analysis of raw data entered either manually at a terminal or acquired on-line during the course of an experiment. Output from the statistical analyses of these experiments, such as mean, standard deviation, statistical significance, and the like, are included in printed reports. In the past, summary information for these results would have to be manually transcribed, then keypunched for entry into the master data base. Because of the extra effort required, some test results were not being entered into the corporate data base, but the original data entries were being preserved on backup tapes. With only paper copy reports available, these experimental results could not be computer searched or readily compared with results of other screening tests.

Diuretic Data. One program area in which data had been recorded on backup tapes but not reported to the corporate research data base was a rat diuretic screen. The original computer programs, written in BASIC and used for data entry and statistical analysis, were also on these tapes. Recorded observations for each experiment were entered into unique data files which would have to be called for by number from within the statistical program. The first line of each of these data files was a literal or header line, an alphanumeric string of variable length, with variable amounts of information entered in any sequence in the line. The possible inconsistency in header line format is not critical if the header is to be used only to identify a printed tabulation of statistical results since the contents can be interpreted by the reader. However, if the contents are to be used to access data and results of analyses for further computer processing, standardization is necessary.

There were three major reasons for writing a set of programs to standardize the data rather than alter the method of data entry. First, it was easier to write two mainline programs, one to handle the data before statistical treatment (the statistical programs were already written), and one to generate reports after the analysis rather than incorporating all three functions into one massive program. Secondly, to enhance the usefulness of the data base, it was essential to include several years of historical information, as well as current information in the master data base. Entering the current data in a new (fixed) format would require either separate programs to interpret the backlog and current data files or re-entry of the header lines for archival data. Thirdly, a fixed format places unnecessary restrictions on the user in that it does not allow as much flexibility for data entry. The additional computing time required for interpretation of the data in a variable format was not considered as important as ease of use for the end user.

To accommodate both current and previous experiments with a single program, an automated system to handle these data was written in three basic modules (Figure 6). Their functions were:

1. Interpretation of file headers and grouping of files by experiment,

2. Statistical analysis of the data, and

3. Generation of reports based on the analyzed data.

Since the statistical programs were already written, module 2 was essentially complete, except for minor modifications to the output of the analysis programs to simplify the processing in module 3. Hence the following discussion deals mainly with modules 1 and 3.

The diuretic data are entered in two files for each test compound or set of control animals. One file, named DATn.BAS (n = 1-999), contains the urine volume and electrolyte assay data for each animal in the treated or control groups (Figure 7). The second file, WGTn.BAS (n = 1-999), contains the individual animal weights. The files are given names in ascending numerical order with the control file receiving the lowest

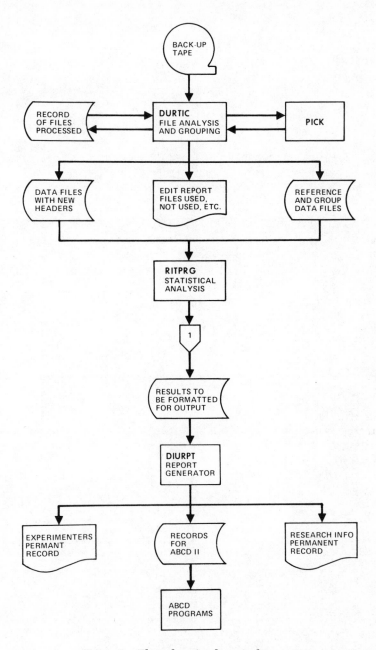

Figure 6. Flow chart for diuretic data system

```
00001 "A-19349 100 MG/KG PO SALINE LOAD 5% M NTR 3-3-77"
00002 8
00003 11.5,114,42,142,2.6
00004 6.8,174,72,217,5.1
00005 9.6,142,57,162,3.9
          ......
00024 9.8,114,80,107,23.7
00025 8.6,122,73,120,17.9
00026 6.9,124,124,79,30.3
00027 END

00001 "CONTROL SALINE LOAD 5% M NTR 3-3-77"
00002 8
00003 2.2,129,188,174,19.9
00004 2,111,64,113,10.3
00005 2.5,106,72,141,12.9
          ......
00024 5.7,159,99,118,35.5
00025 7,194,80,148,38.8
00026 8.2,181,41,148,31.7
00027 END
```

```
00001 19349 A 00100 770303 1 1 2 0 8 3 6 000000000
00002 8
00003 11.5,114,42,142,2.6
00004 6.8,174,72,217,5.1
00005 9.6,142,57,162,3.9
          ......
00024 9.8,114,80,107,23.7
00025 8.6,122,73,120 17.9
00026 6.9,124,124,79,30.3
00027 END

00001 00000 A 00000 770303 0 1 2 0 8 3 6 000000000
00002 8
00003 2.2,129,188,174,19.9
00004 2,111,64,113,10.3
00005 2.5,106,72,141,12.9
          ......
00024 5.7,159,99,118,35.5
00025 7,194,80,148,38.8
00026 8.2,181,41,148,31.7
00027 END
```

Figure 7. Data and control file formats before and after processing by PICK number in a group. The numerical portion of a data file's name and its corresponding weight file are identical.

The header lines of a data file and weight file set also are identical. A batch of test files can be associated with a particular control file by comparing appropriate parameters contained in the header lines of these files.

After some initial manipulating to determine which files are available for a particular processing run, the names of these files are given to the first mainline program, DURTIC. Taking each file in turn, DURTIC uses the file's name and creation date to determine if it was previously processed, i.e., during the last processing run. This is accomplished by comparing the file name and creation date for each new file against the name and date for previously used files. Since the same file name could conceivably be used many times, it is important to include both the name and creation date in this test. If the file was used, DURTIC turns on the "used" indicator for that file and ignores it for the remainder of the run. If the file had not been used previously, subroutine PICK is called to decipher the contents of the header line.

The strategy employed in PICK consists of looking for keywords, phrases, or patterns of characters in the alphanumeric string to determine the value of necessary parameters. The one assumption made was that each variable in the line must be entered in a consistent manner; for example, the compound number must be given as A-19349 not A19349 or 19349-A, etc. All required character strings are specified in the program. However, depending on the nature of the file, whether it contains control data or data on a test compound, some strings will not be searched for. If a header line does not meet all the specified criteria, an error flag is set and the file is ignored during the remainder of the processing run.

Numerous error checks were built into the programs. Only if the data sets passed every test would the experimental results be printed for review and entry into the corporate data base. These checks include:

1. Exact matching of six parameters in the header lines of the data and weight files for each experiment,

2. Exact matching of applicable parameters in the headers for drug and control files for any day's experiments,

3. Matching, within limits, the file's creation date for all data sets in a group,

4. Matching numbers of animals vs. the number of recorded observations, and

5. Matching numbers of observations for control and drug

data sets for each statistical analysis set.

Output from DURTIC consists of an edit report of files used, not used, etc., a file for input to the next run containing the names and creation dates of all files processed this run and not to be reused, and a file containing the names of the files ready to be acted upon in module 2, sorted into appropriate groups.

During the course of the diuretic screening project, several BASIC programs had been used by the investigators to analyze data for the various time periods at which samples were collected. At different times during the screening program the number of chemical assays performed on each sample also varied (e.g. uric acid levels may or may not have been determined). The historical data files do not contain notations as to which time periods or assays are applicable for a particular data set, but these can be ascertained by a count of the actual number of observations that were recorded and the date of the experiment. This is part of the grouping process accomplished in DURTIC.

Following statistical analysis of the data by the appropriate program, the final step in this process (DIURPT) prepares paper reports as well as disk files of the analyzed data formatted for entry into the master data base. Two printed reports and one intermediate data file containing the data formatted for entry to ABCD are generated by DIURPT.

One set of reports, containing the detailed statistical results in tabular form, is used in reporting back to the responsible investigator (Figure 8). The second report is a listing of the data in the format acceptable to ABCD. During the preparation of these reports each experiment is assigned a sequential experiment number by the computer program DIURPT. The last page of each report contains a two-line statement which must be signed and dated by the person who has reviewed the output. This releases the experiments for which results have been printed in the computer output. This same sequence number is incorporated into the data records reformatted for the corporate computer files. Both sets of reports are sent to the pharmacology group for review and signature. Only the copy with the reformatted records is returned to the Research Information group; the responsible investigator keeps the other report on file. Upon receipt of this signed record, the data will be released from the intermediate file to the file being built for the next update of the corporate master data base.

At this time more than 85% of the archival test data processed have been acted upon by the programs without manual intervention. The error checks built into the programs also identify input errors for current experiments. Specific experiments rejected by any of the programs are reviewed by the appropriate individual, corrections made as required, and the data flagged for reprocessing.

SCREEN DU07: T-TEST RESULTS FROM RAT DIURETIC SCREENING   PAGE 9
DATA OFF OF FAILSAFE FOR MONTH OF APR-77   RUN ON 14-FEB-78

	VOLUME (ML/KG)	SODIUM (MEQ/KG)	POTASSIUM (MEQ/KG)	CHLORIDE (MEQ/KG)	NA/K RATIO	URIC ACID (MG/KG)

A-19349   100.00 MG/KG PO SALINE LOAD 5%   M NTR 03-03-77 EXPT #00378

0-2 HOURS

	VOLUME (ML/KG)	SODIUM (MEQ/KG)	POTASSIUM (MEQ/KG)	CHLORIDE (MEQ/KG)	NA/K RATIO	URIC ACID (MG/KG)
CONTROL	10.9682	3.6105	2.5727	4.6805	1.4582	2.8746
DRUG	29.0217	9.9843	3.7212	11.9749	2.6875	2.5400
DELTA	18.0535	6.3738	1.1485	7.2944	1.2293	-0.3345
T-CALC	6.4980	6.5621	2.3864	6.2710	6.2469	-1.3190
PROB	0.0000	0.0000	0.0317	0.0000	0.0000	0.2083

0-6 HOURS

CONTROL	25.3184	8.7457	5.3350	11.2079	1.6776	7.0379
DRUG	59.3786	21.7706	7.2409	25.0231	3.0149	5.9875
DELTA	34.0602	13.0249	1.9059	13.8152	1.3373	-1.0504
T-CALC	11.6807	10.0319	2.9456	10.0495	9.5873	-3.4965
PROB	0.0000	0.0000	0.0106	0.0000	0.0000	0.0036

0-24 HOURS

CONTROL	49.1386	19.7823	10.2726	20.2672	2.0004	26.1265
DRUG	82.1130	29.8380	12.3537	31.9110	2.4113	21.0431
DELTA	32.9744	10.0557	2.0811	11.6438	0.4109	-5.0834
T-CALC	9.7624	6.1872	2.3372	8.1002	2.4598	-3.5757
PROB	0.0000	0.0000	0.0348	0.0000	0.0275	0.0030

2-6 HOURS

CONTROL	14.3502	5.1353	2.7623	6.5274	1.8887	4.1634
DRUG	30.3568	11.7863	3.5197	13.0482	3.3591	3.4475
DELTA	16.0066	6.6510	0.7574	6.5208	1.4705	-0.7159
T-CALC	7.7320	9.5962	2.7917	9.4229	10.5562	-2.2744
PROB	0.0000	0.0000	0.0144	0.0000	0.0000	0.0392

6-24 HOURS

CONTROL	23.8203	11.0365	4.9376	9.0592	2.3939	19.0886
DRUG	22.7345	8.0674	5.1128	6.8879	1.5658	15.0556
DELTA	-1.0858	-2.9691	0.1752	-2.1713	-0.8281	-4.0330
T-CALC	-0.3532	-2.2714	0.2828	-1.9738	-2.4906	-2.8200
PROB	0.7292	0.0394	0.7815	0.0685	0.0259	0.0136

WITH THE EXCEPTIONS, IF ANY, NOTED BELOW EXPERIMENTS #00368 - #00381

ARE OK FOR ENTRY INTO ABCD II _____   _____
                                      SIGNED   DATE

Figure 8.   Sample statistical report

Summary

    The two approaches described in this paper for automation
of test data transmission from the source to a computer master
file have decreased the cost of data entry and increased the
accuracy of data recording. With a totally computerized
operation, changes in experimental parameters and/or statistical
calculations on raw data can be accomodated by modification of
the programs. This allows the scientist flexibility in
recording results of revised experimental procedures.

Abstract

    In order to eliminate, wherever possible, keypunching of
large volumes of data per experiment or transcription of results
obtained from computer statistical calculations, the
experimenters enter original observations directly onto a
computer terminal. All further processing of the data is
controlled by computer programs; humans intervene only when
required for visual evaluation of the information or approval of
outputted results. This approach is cost effective and ensures
producer responsibility for recorded information.
    Two methods are discussed: 1. Direct recording of
observations (e.g. plant growth) on a portable battery powered
recording device, followed by data transmission to and further
processing on the R&D computer; and 2. CRT entry of observed
values for test and control animals for specific tests, followed
by appropriate statistical calculations with report generation
and simultaneous reformatting of summary information for the
corporate master data base (ABCD). Only after verification of
the accuracy of the data by the producers of the information are
the results entered into the master data base.

Acknowledgements

    The authors wish to thank the following for their
assistance and cooperation in these projects: Dr. Amrit Lall,
Patricia Morse, Karen Oheim, Douglas Reno, and Donald Weber.

Literature Cited

1. Morphis, B. B, Torbet, N., Hunter, W. W., and Broome, F. K.,
J. Chem. Doc. (1966) 6, 77-81.

2. Ginsberg, H. F., Greth, P. A., and Morphis, B. B.,"ABCD II -
A User Controlled Biological - Chemical Data System," Presented
before the Division of Chemical Literature, ACS, New York, N.Y.,
August 28, 1972.

3. Federal Register, November 19, 1976, 41 (225), 51206-51229.

4. MSI Data Corporation, 340 Fischer Avenue, Costa Mesa, CA
92627

RECEIVED August 29, 1978.

Clustering in Free-Text Data Bases

RUDOLPH J. MARCUS, EDWIN T. FLORANCE, and EUGENE E. GLOYE

Office of Naval Research, 1030 East Green Street, Pasadena, CA 91106

In previous papers the authors have demonstrated the utility of text handling methods for retrieval and matching of chemical structural information. Although the methods developed have general applicability, the particular data base used was one composed of medical and chemical information and led to heuristic structure-activity correlations.

Specifically, the data base consists of all of the entries from the eighth edition of the *Merck Index* which list a medical use. There are 3,433 such compounds. For each of them we have listed all of the medical uses as well as all of the synonyms by which these compounds are known. In addition to trade names, these synonyms contain one or more Geneva system names. These Geneva system names contain the "structural" information. (The Geneva system of nomenclature has been universally used by chemists since 1864 and is kept up-to-date as new nomenclatural exigencies arise. In the Geneva system, each syllable uniquely defines a structural module of a molecule, and the position of the syllable in the name tells how the modules fit together in the molecule.) It is the manipulation of various Geneva system names as text which constitutes part of the novelty of our work. Material qualifying the medical use terms such as a former use or an experimental use has been coded into the data base. Interesting conclusions from the distribution of such use qualifiers will be listed below. It is seen, then, that our data base is a closed, self-consistent universe which is not updated.

Simple retrieval is possible by searching either by structural module or by medical use. Exhaustive study of the medical use part of the data base began with the counting of uses and the concurrent compilation of an inverted *Merck Index* which could be entered by medical use rather than by chemical compound. In that manner a "sociology" of medical use language was derived (1).

Because both medical use and chemical structure are associated in the same computer file, they form a hyperspace. The nonparametric nature of this hyperspace was discussed in a

previous paper (2) and arises from the natural language text,
rather than numerical, nature (alphanumeric rather than numeric)
of the data base. Clusters in such text systems were defined in
that paper as a collection of vectors whose column elements are
similar. Mutual exclusivity, where the column elements of the
vectors are dissimilar, was also found to be an effective basis
of identifying clusters. While the definition and identification
of clusters in a text data base was satisfying, no quantitative
information about the closeness of elements within the cluster
or about the goodness of fit could be obtained with those definitions. The present paper, therefore, addresses these questions.

Distribution of Uses

The medical use data file which is being analyzed has been
extracted from the main *Merck Index* data file. The use file consists of line records, each made up of three data fields. The
first field is a four-digit code number representing a unique
chemical compound. The second field is a letter character sym-
bolizing a specific qualifier applied to the given medical use.
Table I gives the 10 qualifiers coded in this fashion. Note that
the letter X denotes the set of uses without any qualification.
The third field is the name of the medical use expressed as an
alphanumeric string of characters.

In further manipulation of the use data file, the counting
procedures and concepts of numerical linguistics will be used
extensively. Although this application of linguistics to a data
base not containing the usual natural language text may seem odd,
there are several advantages in doing so. First, the methods
already developed in other contexts may be applied without substantial reprogramming. Second, when the data base is viewed
linguistically, its similarities or differences with other
linguistic data can lead to hypotheses about how the data were
generated. In other words, the rules for naming medical uses
will be compared with the naming rules used in ordinary language.
Third, there is some indication that the problems of indexing
and retrieving can be better understood in this context if the
data base is treated as if it were normal text used for bibliographic purposes; that is, either as key words or as abstracts.

To introduce important concepts, then, it is first necessary to distinguish between a specific use name found in the data
base and all occurrences of that specific name. The specific
name itself will be called a *use type*, while any single occurrence of that name will be called a *use token*. Thus, if the use
narcotic occurs 48 times in the data base, then that set could
be described as 48 use tokens; or, on the other hand, it could
be described as the use type *narcotic* having a frequency of 48
in the data base. This usage is very convenient for counting
purposes.

The data base contains 3433 chemical compounds. There are

TABLE I

SPECIFIC QUALIFIERS GIVEN TO
MEDICAL USES IN THE *MERCK INDEX*

X	Null Character
Z	Additional Information
H	Has been used
F	Formerly
A	Activity, Properties
E	Experimental
I	Has been investigated, has been studied, investigative, has been tried as
P	Proposed as
R	Reportedly causes
S	Occasionally as, sometimes as

949 distinct medical uses, or use types, mentioned in the use file. But since each compound may have more than one different use, the total number of use tokens is expected to be larger than the total number of compounds. In fact, there are 4848 use tokens in the file, which represents about 1.4 uses on the average for a typical compound. Hence, each use type occurs on the average of about 5.1 times; that is, there are over 5 use tokens for each typical use type.

Statistics like those just cited--average number of uses per compound or average use tokens per type--apply to the entire data base and give some notion of gross properties. But more detailed structures can be measured and will give a better picture of the naming rules which led to assignment of use names. In fact, more can be learned by partitioning the data set into smaller subsets and calculating the properties of these divisions. To the extent that these subsets differ substantially from the whole file, it is possible to say that their selection represents a significant analytical operation on the data.

Use Combinations. Each compound in the data base has one or more distinct use names associated with it. For some compounds, the set of names may contain two that are the same but whose (letter) qualifiers are different. These differently qualified use names will be considered as different use names for the purpose of this investigation. Thus, each compound can be assigned an integer which denotes how many distinct use tokens are associated with it. This integer, which measures uses per compound, can also be assigned to each use token associated with the given compound. In the data base, the number of uses per compound varies generally up to 6, but there are a special set of compounds which have 9 associated uses each.

The first restructuring of the data file consisted, then, in assigning a uses-per-compound integer to all use tokens and an assignment of a code number to each use type. This latter assignment was made by letting the most frequent use type have the lowest code number. The data file was further reorganized by combining all use tokens (now represented by code numbers) for a given compound on the same line in ascending numerical order. Thus, the 4848-line file was reduced to a 3433-line file. Only the use qualifier data was removed. The resulting file was then sorted by uses per compound and secondarily by the particular combination of uses. Finally, a counting program was used to eliminate compound numbers and group together all similar use combinations.

This partitioning of the main file into subsets with different values of uses per compound (that is, size of the use combination set) provided some statistics on how multiple uses are distributed. Table II presents the summary data. It shows how the number of compounds and use tokens are distributed over the uses/compound parameter. The column labeled "use types"

TABLE II

MULTIPLE USE TABLE SUMMARY DATA

Distribution of Number of Compounds and Their
Occurrence Over the Uses/Compound Parameter

Uses/ Compound	Number Of Compounds	Use Tokens (Occurrences)	Use Types (Descriptors)	Distinct Use Combinations
1	2447	2447	570	570
2	717	1434	414	391
3	187	561	231	133
4	60	240	140	46
5	5	25	21	5
6	4	24	20	4
9	13	117	10	2
TOTALS	3433	4848		1151

gives the number of use types which appear in each subset. Since
the partitioning is overlapping in use types, the numbers in that
column cannot be added to give a total. The last column indicates how many distinctly different use combinations there are
with a given combination size. For example, each compound with
5 or 6 uses has a distinct set of associated uses, while 9-use
compounds have only 2 distinct sets. Since use combinations are
partitioned in a non-overlapping fashion, a sum of the column
entries gives 1151 as the total distinguishable number of use
combinations. Thus, on the average, each use combination is
associated with about 3 different compounds.

In one version of the file sorted on use combination, all
use sets beginning with the same use are grouped together, regardless of the size of different combinations. This format permits
comparison of use combinations containing the same pairs of uses.
An eyeball search, with some computer assistance, was made to
tabulate the most frequent use pairs occurring in the use file.

Table III includes all use pairs with a frequency of occurrence greater than 5. The pair names and frequency of occurrence
are presented in the table, as well as the distribution of these
pairs over different size combinations. For example, the use
pair *analgesic-sedative* occurs three times as a pair but also in
three triple combinations with other uses: hypnotic, narcotic,
and skeletal muscle relaxant.

The entries in Table III have been grouped by closeness of
relation. In a taxonomic sense, the grouped parts constitute a
cluster. The first eight entries represent an analgesic-sedative
cluster, which includes the use names *analgesic* or *sedative*. The
next four table entries represent an antiseptic-astringent
cluster. The *cardiotonic-cardiac rate decrease* pair is isolated
and seems to suggest synonymity. The next five entries are a
cluster centered on the *diuretic-antihypertensive* pair.

Proceeding further down the table, one encounters a cluster
based on a triple of uses: adrenocortical steroid, glucocorti-
coid, anti-inflammatory. The *parasympathomimetic-miotic* cluster
occurs alone, followed by a cluster centered on *sympathomimetic*.

Use-Oriented Data. To investigate related uses, the data
base was restructured to place all uses associated with a given
compound together in the same record (or equivalently on the same
text line). But other information may be obtained by grouping
together all tokens of the same use type. In a file whose
records consist of the compound code, the uses-per-compound
parameter, the qualifier code, and the use name, a simple
alphanumeric sort on use name brings together all use tokens
of the same type. The resulting file can then be further sorted
on either uses per compound or qualifier code. From these sorted
files, which still contain compound codes, a summary table can
be prepared by removing compound references. In such a table,
each record or table line contains a count of all use tokens

TABLE III

DOMINANT USE PAIRS

Uses/Compound				Freq.	Pair Names
2	3	4	≥5		
29	12	1	14	56	analgesic, antipyretic
27	5	1		33	analgesic, narcotic
			10	10	analgesic, anesthetic
3	5		1	9	analgesic, antirheumatic
4	4	1		9	analgesic, antitussive
6	1			7	analgesic, antispasmodic
3	3			6	analgesic, sedative
54	3	1		58	sedative, hypnotic
4	7			11	sedative, tranquilizer
14	12			26	antiseptic, astringent
12	1			13	antiseptic, disinfectant
4	1		1	6	antiseptic, expectorant
6	2			8	astringent, styptic
22				22	cardiotonic, cardiac rate decrease
	15	7		22	diuretic, smooth muscle relaxant
	14	7		21	diuretic, myocardial stimulant
14				14	diuretic, antihypertensive
		7		7	diuretic, vasodilator
10	2			12	antihypertensive, ganglion blocking agent
8	6	1		15	adrenocortical steroid, glucocorticoid
4	6	1		11	adrenocortical steroid, anti-inflammatory
5	2	1		8	parasympathomimetic, miotic
4	2			6	sympathomimetic, vasoconstrictor
5	1			6	sympathomimetic, CNS stimulant
4	2			6	sympathomimetic, decongestant

(occurrences) of a given type (the frequency of that type) together with the use name. Separate columns then present counts of the numbers of use tokens appropriate to each value of the secondary sort parameter. Table IV presents a sample of such a tabular presentation for uses per compound. Note that the table has been sorted so that the most frequent uses occur first in the table.

The type of data file displayed in Table IV resembles a concordance in the sense that it shows occurrences of different words in different subclasses or texts. In most linguistic studies, the columns represent different texts, and the table entries denote counts of word tokens in those texts. We will use the description "text" here to denote one of these columns even though no text has been defined; rather, "text" means a partitioning of the data. The entire table will be called a word distribution table, or use distribution file, when referring to these particular data.

An inspection of Table IV reveals that there appear to be two distinctively different classes of use types as revealed by the distribution of tokens with the parameter uses/compound. A comparison of *antimicrobial* with *analgesic* is instructive. *Antimicrobial* has the largest number of single uses (201) in the file and has only 25 multiple uses. It also describes no compounds with more than three uses. This suggests either that compounds with antimicrobial properties are highly specific or that the name is applied very specifically. In contrast, *analgesic* has only 55 single uses, but 119 multiple uses. In fact, one compound with analgesic properties has six uses. This suggests either that compounds with analgesic properties are rather nonspecific and generalized in their action or that the name *analgesic* refers to a rather wide range of effects.

These examples appear to suggest that, by examining the distribution of tokens with the uses-per-compound parameter, the uses can be either partitioned into specific and nonspecific classes or described by a specificity parameter.

Frequency Distribution. The word distribution table provides not only the distribution of word tokens among the various texts but also the total frequency of each word type. The total number of types or tokens having a given frequency can be determined by simply counting types in the distribution table or summing the total number of tokens with that frequency. The functions which present the number of types or tokens of a given frequency are called, respectively, the *type* and *token distribution functions*.

In general word statistics, the type frequency distribution is a steeply decreasing function of frequency; the singly-occurring words constitute between 25% and 40% of the total number of types. At frequencies greater than 30 or 40 (for a small corpus), the type distribution function is zero at many

TABLE IV

USE DISTRIBUTION FILE

Frequency Sort for Uses Per Compound

	Uses/Compound						Freq.	Use Name
1	2	3	4	5	6	9		
201	24	1					226	antimicrobial
93	55	24	1	1	1		175	antiseptic
55	82	21	2		1	13	174	analgesic
30	72	15	3				120	sedative
47	36	22	9	2	1		117	diuretic
54	23	7	2	1			87	antispasmodic
22	59	4	2				87	hypnotic
71	10	1					82	antineoplastic
52	10	3	2		2	10	79	anesthetic
46	20	7					73	tranquilizer
58	10	2					70	antihistaminic
19	32	17	1				69	astringent
5	33	12	1		2	13	66	antipyretic
27	32	3					62	antihypertensive
13	22	12	6	2	1		56	expectorant
14	29	4	4				51	sympathomimetic
29	16	5					50	anthelmintic
15	20	8	3	2			48	carminative
12	29	6	1				48	narcotic
23	8	1	1			13	46	antimalarial
29	11	4	1				45	antitussive
35	8	1					44	antituberculous
40	1						41	x-ray contrast medium
27	5	4	3				39	cathartic
23	10	5	1				39	CNS stimulant
32	1	2				3	38	anesthetic local
12	24	1	1				38	cardiotonic
16	7	6	9				38	vasodilator
10	13	6	2	2	2		35	counterirritant
23	9	2		1			35	laxative
12	12	6	1				31	adrenocortical steroid
5	3	16	7				31	smooth muscle relaxant
22	7	1					30	amebicide
14	14	2					30	ganglion blocking agent
14	7	4	2				27	parasympathomimetic
21	5						26	vasodilator coronary
20	3	1					24	antifungal
23	1						24	estrogenic
2	1	14	7				24	myocardial stimulant

frequencies and has values between 1 and 5. At large frequencies
the function is zero except at a very few frequencies. The
nature of token frequency distribution functions is somewhat
different, even though for large frequencies they are also non-
zero at the same selected frequencies. For purposes of analysis
and modeling, it is better to work with cumulative distributions.
Cumulative distributions are formed by summing the frequency
functions for all frequencies less than a certain value and
dividing by the total sum. Cumulative distributions, when
plotted as a function of frequency, are S-shaped (ogive) curves,
ranging from 0 at zero frequency to 1 at large frequency.

Numerical linguists have made many attempts to describe or
model the word frequency distribution. One of the models which
has extensive use is the log-normal distribution. This model can
be defined in terms of the cumulative normal probability distribution. The log-normal distribution assumes that when the cumulative distribution is plotted versus the logarithm of frequency,
the resulting curve conforms to a normal distribution. The log-
normal distribution describes many long-tailed distributions in
linguistic science; in particular, the distributions of word
length and sentence length (3). Log-normal distributions are
also found in econometrics and other social phenomena (4). What
gives the log-normal distribution particular interest in word
frequency analyses is that if the type frequency distribution is
log-normal, then the token frequency distribution is also
log-normal.

In practical terms, the type and token cumulative frequency
distributions may be tested for log-normality by plotting these
functions on normal probability paper with the logarithm of frequency as the abscissa. When this test was applied to the
medical use type and token frequency distributions, the log-normal
model was found to describe both distributions very well over
two orders of magnitude. Figure 1 shows this relationship for
the use type distribution and also for the cluster type distribution. The cluster types involve a different counting than use
types, so that two clusters are the same if all their components
are identical. The total number of cluster types is simply the
sum of the last column in Table II.

Plotted for comparison on Figure 1 are two cumulative distributions in which the frequency is distributed normally. Note
that the normal curves cannot satisfactorily fit the long tails
of the log-normal distributions. While the normal curves have
been chosen somewhat arbitrarily to fit the observed distributions at frequencies of 1 and 22, there is no consistent way in
which to force a logarithmic curve to fit a linear one.

Plotted in Figure 2 are the cumulative type frequency distributions for two subsets of the data base. The single use
curve corresponds to all compounds with a single use. The X
curve corresponds to all uses with the X (null) qualifier; i.e.,
all uses which are unqualified. These subsets are also seen to

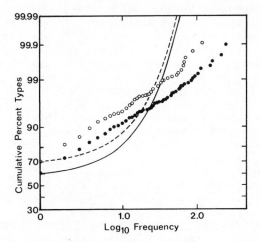

Figure 1. Log-normality of type distribution. Points are plotted on a log_{10} frequency scale; curves represent a normal, rather than log-normal, distribution in frequency. Cluster types, ○; use types, ●. Normal curve vs. frequency fit to data: clusters, – – –; uses, ——.

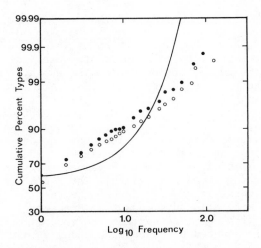

Figure 2. Log-normality of data base subsets. Points are plotted on a log_{10} frequency scale; curve represents a normal, rather than log-normal, distribution in frequency. Single uses, ●; X uses (no qualifier), ○. Normal curve vs. frequency fit to single-use data.

be log-normally distributed.

The fact that the use frequency distribution conforms so closely to the log-normal was unexpected, inasmuch as the statistical "rules" which operate in natural language text would not necessarily be assumed to operate for a corpus of terms derived from a highly technical area of specialty such as pharmacology. However, the finding suggests that the *Merck Index* medical use data base does not differ substantially in its linguistic properties from natural language data bases. This result is consistent with studies of disease identification, which also involves a professional code of terms (3).

Specificity. Several attempts were made to develop a nonparametric statistical description of the specificity question, based on techniques from numerical linguistics. It was hoped that sets of very specific or very nonspecific uses would naturally emerge in the analysis. But it was found that a simple parametric statistic did far better than nonparametric ones: the mean uses per compound for each use. The mean uses per compound for the entire data base (excluding 13 compounds with 9 uses, which represent a special case) is calculated to be 1.739. The difference between the population mean and the mean for any use can be tested for significance by comparison to the standard deviation. For a large use frequency, N, the mean uses per compound, $\bar{u}$, is distributed normally with a standard deviation given by

$$\sigma = (.899/N)^{\frac{1}{2}}$$

Figure 3 shows the distribution of values of $\bar{u}$ for frequencies of 6 or greater. The curves representing $\pm 2\sigma$ and $\pm 3\sigma$ limits about the population mean are also shown. From this figure, 25 uses were found for which $\bar{u}$ differed by 3σ from the population mean. Eight uses have significantly low $\bar{u}$: seventeen uses have significantly high $\bar{u}$. Lists of these uses with values of $\bar{u}$ and standard deviations are given in Table V.

The parameter $\bar{u}$ was also calculated for the set of compounds corresponding to each use qualifier. The sets described by qualifiers X and Z both showed $\bar{u}$ values not significantly different from the total population. The H and F sets both showed higher than average $\bar{u}$. Except for the R qualifier, the remaining sets all had $\bar{u}$ less than average. Table VI shows these results with combined sets to increase significance. Significance is denoted by calculated σ deviations from the population mean. In this table, X and Z are average, H and F combined have significantly higher uses/compound, and the remaining combination has significantly lower uses/compound.

Thus X and Z compounds seem typical of the average compound in the data base, while a compound with an H or F qualifier has significantly more uses than average. This result suggests

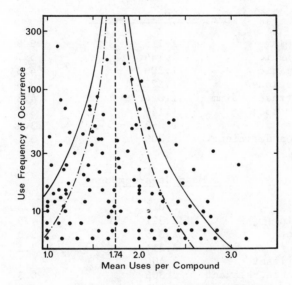

*Figure 3.   Distribution of mean uses per compound.
Outliers are identified in Table V. Population mean,
– – –; 2 σ deviation, – · – ·; 3 σ deviation, ———.*

TABLE V

USES HAVING SIGNIFICANT SPECIFICITY PARAMETER

Use Name	Mean Uses Per Compound	Deviation From Population Mean (In σ's)
Low Mean Uses/Compound		
antimicrobial	1.115	9.9
antineoplastic	1.146	5.7
antihistaminic	1.200	4.8
antituberculous	1.227	3.6
x-ray contrast medium	1.024	4.8
anesthetic local	1.143	3.7
estrogenic	1.042	3.6
anemia iron deficiency	1.000	3.1
High Mean Uses/Compound		
diuretic	2.026	3.3
antipyretic	2.321	4.5
expectorant	2.375	5.0
vasodilator	2.211	3.1
counterirritant	2.400	4.1
smooth muscle relaxant	2.806	6.3
myocardial stimulant	3.083	6.9
antirheumatic	2.591	4.2
glucocorticoid	2.500	3.3
mydriatic	2.750	3.7
CNS depressant	2.700	3.2
decongestant	2.700	3.2
diaphoretic	2.700	3.2
carbonic anhydrase inhibitor	2.778	3.3
dermatides	2.857	3.1
dermatoses	3.167	3.7
sudorific	3.167	3.7

either that compounds with many uses tend to have more "old" uses or that "old" use names tended to be less specific. The remaining qualifier group has significantly fewer uses than average, which suggests either that compounds with "new" medical uses tend to be labeled with only one use or that "new" descriptors are more specific.

Discussion

We have described in the preceeding section four different cuts through the hyperspace of medical uses and their frequency in the *Merck Index* data base. By "cut" we refer by analogy to a projection of the hyperspace onto one or more recognizable orthogonal Cartesian axes. In this section we will discuss what we have learned from these four cuts. We will also approach, but certainly not exhaust, the association of this hyperspace (the effect space) with the hyperspace containing the chemical structure of compounds having these uses (the structure space).

Use Frequencies. One cut through the data base is to plot the frequency distribution of the medical uses and various major subsets thereof. We show in Figures 1 and 2 that the cumulative distribution of both use types (descriptors) and use tokens (occurrances) is log-normal. Two major subsets of the data base also have a log-normal distribution. This type of distribution shows that the vocabulary of this unique text is, contrary to what might have been expected, no different from any other text which has been considered by numerical linguists. The log-normal distribution found by numerical linguists applies not only to vocabularies involving disease identification, but applies to a number of different natural language data bases.

Mean Uses/Compound. Another cut through the data base has been made in terms of mean uses per compound. This parameter has been plotted against frequency in Figure 3. It is this figure which permits a first statistical evaluation of the significance of clusters occurring in text data bases. Lines corresponding to two and three standard deviations (σ) from the mean are drawn into Figure 3, and the various outliers on either side are listed in Table V. The reasons why particular uses have low or high mean uses/compound are perhaps best left to pharmacologists. It is readily apparent, however, even to nonpharmacologists, that the uses listed under low mean uses/compound are highly specific in a linguistic, as well as statistical, sense. Four of them are "anti-," and the other stem of these four words is a highly specific concept: -microbial, -neoplastic, -histaminic, and -tuberculous. Another three uses under low mean uses/compound depend on distinctly chemical properties; these are x-ray contrast medium, anemia iron deficiency, and estrogenic. Uses listed under high mean uses/compound seem to be of two kinds. One of

these is the kind of use such as diuretic, vasodilator, smooth
muscle relaxant, myocardial stimulant, antipyretic, and expec-
torant, which are present in the dominant use pairs, or clusters,
listed in Table III and which will be discussed later on in this
section. Obviously, such compounds would, by their presence in
these clusters, be expected to have more uses per compound than
the average. Other uses in this high mean uses/compound rubric
are those use designations which are older and perhaps somewhat
imprecise by more modern definitions. This point will become
apparent in the next paragraph when we discuss qualifiers.

　　Qualifiers. A third cut through the hyperspace was taken
on the basis of the qualifiers which are defined in Table I. It
was shown in Figure 2 that a major subset defined by the use
qualifier X (a null character) essentially followed the same log
normal distribution as did the entire data base of medical uses.
Table VI shows that both X and Z ("additional information") subsets do not have a statistically significant deviation from the
population mean. However, uses having qualifiers H or F, which
represent former uses, have highly significant (5.3 σ) greater
mean uses/compound, a fact which was alluded to in a different
context in the previous paragraph. Fortunately for this statistical view of the *Merck Index* data base, the use qualifiers A,
E, I, P, R, and S, which represent experimental compounds not
yet in general use, together have fewer mean uses/compound with
statistically high significance (4.5 σ). The discussion in this
section indicates that not all subsets of the data base have the
same distribution and that use qualifiers, such as those listed
in Table I, which were frequently used in the eighth edition of
the *Merck Index*, can distinguish between different kinds of medical uses in a statistically significant manner.

　　Use Correlations. A fourth cut through the medical use-
frequency hyperspace is the one which involves the derivation
of dominant use pairs. Table III lists these use pairs as a
function of uses/compound. Table III also lists the frequency
of individual use pairs. It was noted earlier that the use pairs
in Table III have been grouped by closeness of relation and that
in a taxonomic sense, the grouped parts constitute a cluster.
The seven clusters identified are shown in Table III as separated groups of use pairs. It is this particular cut through
the data base which suggests the possibility of extrapolation.
Many compounds which have one use contained in a cluster might
well be examined pharmacologically or clinically to see whether
that compound also has one or more of the other uses in that
cluster.
　　The same reasoning which suggests this possibility of
extrapolation also leads to the conclusion that it is this
particular cut through the medical use hyperspace which leads
into associations with the chemical structure hyperspace.

TABLE VI

QUALIFIER SUBSETS OF USE TOKENS

Use Qualifier*	No. of Uses	Mean Uses/Compound	Deviation From Population Mean (In σ's)
X	2308	1.711	1.5
Z	1062	1.711	1.0
H + F	1106	1.881	5.3
A + E + I + P + R + S	255	1.490	4.5
Total Uses**	4731	1.739	

* Definitions in Table 1.
** No 9-use compounds included.

Ideally, each of the uses in Table III can be associated with a particular chemical structure. In previous work (1) we have shown that such associations can be made not only by pharmacologists, working from laboratory data and current theories, but also by text manipulation of the kind which we have described here. Thus for example, the effects of compounds in the *sympathomimetic* cluster have been ascribed to the presence of the ethylamine grouping, while the effect of compounds in the *parasympathomimetic* cluster have been ascribed to compounds containing the ammonium ion structure or its phosphorus analog. The compounds in the *adrenocortical steroid* cluster have similar chemical structures by definition. The *cardiotonic* cluster contains natural products related to those found in digitalis and containing similar side chain structures. The *antiseptic-astringent* cluster contains recognizably caustic compounds. It is hoped that similar associations can be made in the *diuretic* and *analgesic* clusters. The latter, in particular, has the largest number of uses and perhaps those with the greatest variety of definitions. It would be interesting to be able to resolve these definitions on the basis of chemical structure as well as numerical linguistics.

Abstract

    Heuristic clustering methods for text data have been applied to a data base describing chemical compounds with medical uses. Clusters are sets of chemical compounds related by similarity of both chemical structure and activity. Previously discovered clusters in a free-text data base taken from the *Merck Index* could be described by a 2 x 2 activity-structure matrix. A statistical examination of the activity descriptors (medical uses) reveals them to have a log-normal distribution over two frequency decades. Log-normal distributions have been found by workers in other disciplines to be characteristic of random selection from a set of items with fixed limits. An investigation of compounds with more than one medical use revealed that the dominant clusters are labeled with the descriptors *analgesic/sedative, antiseptic/astringent, diuretic/antihypertensive, adrenocortical steroid, parasympathomimetic,* and *sympathomimetic.* Chemical similarities are involved in those clusters based on multiple medical uses as well as in the case of clusters based on the 2 x 2 activity-structure matrix, which involved computer searches on single activity descriptors only. Consideration of multiple medical use clustering assures comprehensiveness and supplements any weakness of purely heuristic searching. The existence of these chemical similarities permits the exploration of new or previously unreported uses of chemical compounds by computer manipulation of text data.

Literature Cited

1. Marcus, R. J. and Gloye, E. E., "Real-Time Interrogation of Chemical Data," *J. Chemical Documentation* 11, 163-7 (1971).

2. Marcus, R. J., Gloye, E. E., and Florance, E. T., "Computer Search of a Free-Text Data Base as a Tool for Investigating Structure-Effect Relationships," *Computers and Chemistry* 1, 235-241 (1977).

3. Herdan, G., "The Advanced Theory of Language as Choice and Chance," Springer-Verlag, New York, N.Y., 1966.

4. Montroll, E. W. and Badger, L. W., "Introduction to Quantitative Aspects of Social Phenomena," pp. 110-120, Gordon and Breach, New York, N.Y., 1975.

RECEIVED August 29, 1978.

5

CHEMLINE: a Chemical Structure Search Key to Biological Information

MELVIN L. SPANN, DONALD J. HUMMEL, ROBERT J. SCHULTHEISZ, SHARON L. VALLEY, and DONALD F. WALKER, JR.

Toxicology Information Program, National Library of Medicine, Bethesda, MD 20014

This paper demonstrates the ability of an on-line chemical dictionary file in accessing, through chemical descriptors, computerized files containing biological data and information. As indicated in Figure 1, CHEMLINE, the National Library of Medicine's (NLM) interactive, on-line dictionary file, can be viewed as the focal point for the chemical searching of the Library's on-line literature retrieval services (1). In addition, CHEMLINE provides a linkage through a "Locator" designation to other files containing information relevant to health and environmental concerns. An example of the latter is the Environmental Protection Agency's Toxic Substances Control Act (TSCA) Inventory Candidate List. Substances appearing on the Candidate List are referenced with the EPATSCALIST Locator code. This discussion will be limited to CHEMLINE's use in connection with the TOXLINE and RTECS files, which are available to over 800 institutions that subscribe to NLM's on-line services.

CHEMLINE

CHEMLINE (CHEMical dictionary on-LINE) is a file of chemical descriptors created by NLM's Toxicology Information Program in collaboration with Chemical Abstracts Service (CAS). This file contains nearly 500,000 chemical substance names representing over 246,000 unique substances. Because of CHEMLINE's unique file design, it has capabilities which support both full structure and substructure searching.

Figure 2 shows a typical CHEMLINE unit record. The following section highlights the characteristics of the fields within this record, all of which are directly searchable (up to 39 characters) and printable.

RN is the Chemical Abstracts Service (CAS) Registry Number. This uniquely assigned number of up to nine digits appears in a hyphenated format without leading zeros.

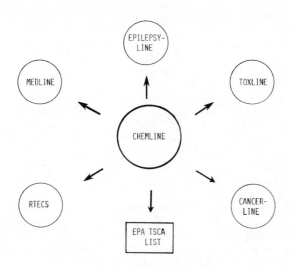

Figure 1.   NLM's CHEMLINE file

```
RN  - 553-24-2
MF  - C15-H16-N4.CL-H
N1  - C.I. BASIC RED 5, MONOHYDROCHLORIDE (8CI)
N1  - 2,8-PHENAZINEDIAMINE, N(8),N(8),3-TRIMETHYL-,MONOHYDROCHLORIDE
      (9CI)
SY  - C.I. 50040
SY  - AMINODIMETHYLAMINOTOLUAMINOZINE HYDROCHLORIDE
SY  - NEUTRAL RED
SY  - TOLUYLENE RED
SY  - NEUTRAL RED W
SY  - C.I. BASIC RED 5
SY  - NUCLEAR FAST RED (BASIC DYE)
WL  - T C666 BN INJ E FZ LN1&1 &GH
NR  - 3
RS  - 6,6,6
RE  - C4N2-C6-C6
CL  - NC2NC2
MH  - PHENAZINES / (69-74)
MH  - DIMETHYLAMINES (73-74)
MH  - NEUTRAL RED / (MINOR IH 75)
LO  - TOXLINE
LO  - MEDLARS
LO  - EPATSCALIST
RC  - R089-4926
```

Figure 2.   CHEMLINE record

MF is the molecular formula of the compound expressed in the Hill convention. This means that, for organic compounds, the number of carbon atoms is cited first; this is followed by the number of hydrogen atoms; and then all other elements are cited in alphabetical order. For inorganic compounds, all elements occur in alphabetical order.

N1 is the Chemical Abstracts (CA) preferred index name, i.e., the systematic names used in the Chemical Substance and Formula Indexes of CA. The 8CI and 9CI designations following the two N1 names for this record indicate that the name comes from the Eighth Collective Index Period (1967–71) and the Ninth Collective Index Period (1972–76) of CA. One can see from this example that the 9CI nomenclature is chemically more standardized than the 8CI nomenclature.

SY indicates the synonymous names that CAS has on record for a chemical substance. Within the synonym field will be found some uninverted chemical names, generic, trivial, trade and experimental names, as well as company code numbers for chemicals.

WL is the Wiswesser Line Notation (WLN) field. The WLN is a unique and unambiguous representation of a chemical structure diagram using a linear arrangement of 36 alphanumeric characters, 3 special characters and the blank. There are approximately 10,000 CHEMLINE records that contain WLNs.

The next set of fields (NR through CL) in this record contains ring information. Since about 80% of the substances within CHEMLINE contain ring systems, one can see the importance of incorporating this information into a chemical structure search system, especially for substructure searching.

NR provides the number of component rings within each unique ring system in a chemical substance. As seen in the diagram, there are 3 rings within the phenazine ring system.

RS gives the size of the component rings within each unique ring system. For multiple ring systems, ring sizes are cited from smallest to largest. In the phenazine ring, the ring sizes are 6,6,6.

RE, the Ring Elemental analysis field, contains the molecular formula for each component ring within a unique ring system. In this field, the element count begins with carbon and all other elements (excluding hydrogen) follow in alphabetical order. For a multiple-ring system, the ring elemental analysis is given first in order of ring size and secondly in ascending order of the carbon atom count for individual rings when ring sizes are equal. Therefore, the RE field for Basic Red 5 is C4N2–C6–C6. Note that this field is alphabetical and does not necessarily provide the order of occurrence of the component rings within a ring system.

CL is the Component Line formula field which provides a
topological description of those rings containing between 4 and 8
atoms that also contain two or more non-carbon atoms.   The rules
for deriving a Component Line formula are as follows:

1. Start at the earliest alphabetic non-carbon atom;
2. proceed around the ring in the direction that
    provides the shortest path to the next non-carbon
    atom; and
3. where contiguous atoms repeat, give only the
    atom and its number of occurrences.

Thus, the example shown has the Component Line formula NC2NC2.

MH or the MeSH heading field contains terms from NLM's
Medical Subject Headings (MeSH) controlled vocabulary.   There
are approximately 4,000 CHEMLINE records with MeSH headings.

LO contains the LOCATOR designations which identify sources
of information or citations relevant to the specific chemical
substance retrieved.   CHEMLINE presently contains locators to
MEDLARS, TOXLINE and EPATSCALIST.   A locator to RTECS will soon
be added.

RC contains the alphanumeric code assigned to the approximately 33,000 compounds appearing on the EPA TSCA Inventory
Candidate List of Chemical Substances.   This code should be
reported to EPA, along with the CAS Registry Number, to satisfy
TSCA reporting requirements.

In addition to the information fields listed in Figure 2,
there are several searchable but non-printable fields within
CHEMLINE records that do not appear in this figure.   Only two of
these fields which are particularly useful in performing substructure searches will be discussed.   The first is the NF or
Name Fragment field.   Name fragments are generated by computer
programs from the N1 (Type 1 name) and SY (Synonym) fields by
breaking a name on hyphens, colons, enclosures and blanks.
Each uniquely occurring character string becomes a searchable
entity.   Thus, the name fragments for the second N1 name shown
in Figure 2 are:

2,8	3
phenazinediamine	Trimethyl
N	Monohydrochloride
8	

The contents of the second fragment field, FF, are the
molecular formula fragments which are derived by breaking the
molecular formula on hyphens and periods.   Excluding hydrogen
atoms, each elemental symbol and its count are stored as searchable entities.   In addition, the elemental symbol without a count
is saved for the heteroatoms (nitrogen, oxygen, phosphorus and
sulfur) and the halogens (fluorine, chlorine, bromine and iodine).

This feature permits one to search for any (unspecified) number
of hetero or halogen atoms in a chemical compound, or for a
specific number of these atoms. As an example, searching with
the formula fragment S would retrieve all substances having at
least one sulfur atom in its structure; whereas formula
fragment S1 would retrieve records for substances having only one
sulfur atom. The molecular formula fragments retained for
Basic Red 5 are:

> C15
> N4   N
> CL1   CL

TOXLINE

     TOXLINE (TOXicology Information on-LINE) is an extensive
collection of computerized toxicology information with over
425,000 references (as of 1978) to published human and animal
toxicity studies, effects of environmental chemicals and pollutants, and adverse drug reactions. TOXLINE covers the published
literature from 1974 forward. Older information (380,000 references) can be found in TOXBACK, which is searchable in an off-
line mode.
     TOXLINE/TOXBACK are made up of component files that come
from the major secondary literature sources, as well as specialized sources. The collection of bibliographic citations are
obtained either through profiling certain files or selecting complete specialty files. The sources and extent of coverage of
component files are: Chemical-Biological Activities (CBAC)
Sections 1-5 from 1965, Sections 62-64 from 1975 and Sections 8,
59 and 60 from 1975; Toxicity Bibliography (TOXBIB) from 1968;
Abstracts on Health Effects of Environmental Pollutants (HEEP)
from 1972; International Pharmaceutical Abstracts (IPA) from 1970;
Pesticides Abstracts (PESTAB, formerly HAPAB) from 1966;
Environmental Mutagen Information Center (EMIC) file from 1960;
Environmental Teratology Information Center (ETIC) from 1950;
and the Toxic Materials Information Center (TMIC) from 1971-1975.
In addition, TOXBACK contains a teratology file (TERA) covering
the years 1960-1974 and a special collection of literature concerning the health effects of pesticides, the Hayes File, which
covers the period from 1940-1966.
     As indicated in Figure 3, all of these files are arranged so
that they can be searched simultaneously in response to a search
query; they can also be 'searched independently through use of the
file acronym. Each record in TOXLINE contains a full bibliographic citation, most have abstracts and/or indexing terms and CAS
Registry Numbers. On-line retrieval of TOXLINE records is
usually accomplished by free-text searching of TEXTWORDS (TW) or

keywords found in titles, index fields or abstracts of citations. Other searchable elements include author name, CAS Registry Number, secondary source identification, language, year of publication, and journal coden.

RTECS

While TOXLINE/TOXBACK are bibliographic retrieval services, the third file to be briefly described in this paper can be considered an on-line _data_ retrieval file. RTECS (Registry of Toxic Effects of Chemical Substances) is a product of the National Institute for Occupational Safety and Health (NIOSH) and is available as an annual publication (2). NLM obtains a computer-readable file from NIOSH who maintains responsibility for file contents. The present (1977) version of RTECS contains toxicity data for approximately 26,000 substances. Table 1 provides a description of an RTECS unit record.

The RTECS fields identified with a search abbreviation are directly searchable (up to 36 contiguous characters). The information in the Toxicity Data Index Strings (IX) is linked to prevent false associations and is structured as follows:

ROUTE;SPECIES;ORDER;STUDY TYPE;VALUE;
TOXIC EFFECTS (if any);SPECIFIC EFFECT/ORGAN SYSTEM AFFECTED

An example of this categorization would be:

ORAL;RAT;RODENTS;LD50;54 MG/KG;TOXIC EFFECTS;CARCINOGENIC

Each term or phrase between semicolons is directly searchable and can be ANDed together. For example,

(IX) ORAL AND RAT AND LD50 AND CARCINOGENIC

The information contained in the fields that are asterisked in Table 1 are searchable as free text terms; that is, unique words which can be ANDed or ORed together. The (CT) field identifier is used when searching the Toxic Data Source field and (TW) is used when searching for information in the remaining "free text" fields.

CHEMLINE-TOXLINE Search

The previous discussions presented an overview of three of the National Library of Medicine's on-line files; next will be an examination of the various ways in which CHEMLINE can be utilized in enhancing the chemical accessibility to toxicological information in another file. Using the pesticide, Leptophos, as an example, Figures 4 and 5 show how this is done.

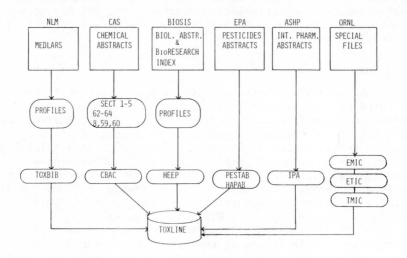

Figure 3.   Sources of TOXLINE bibliographic files

```
SS 1 /C?
USER:
"FILE CHEMLINE
PROG:
YOU ARE NOW CONNECTED TO THE CHEMLINE FILE.

SS 1 /C?
USER:
(SY) LEPTOPHOS
PROG:
SS (1) PSTG (1)

SS 2 /C?
USER:
"PRT RN,SY,LO
PROG:

1
RN  - 21609-90-5
SY  - VCS 506
SY  - VELSICOL VCS 506
SY  - VELSICOL 506
SY  - PHOSVEL
SY  - K62-105
SY  - LEPTOPHOS
SY  - FOSVEL
SY  - NK 711
SY  - ABAR
LO  - TOXLINE
LO  - MEDLARS
LO  - EPATSCALIST
```

Figure 4.   CHEMLINE synonym
search

Searching the synonym field of CHEMLINE for Leptophos retrieves one record. PRINTing the RN, SY and LO fields provides the CAS Registry Number, other synonyms which are known for the compound and the source of additional information.

As seen in Figure 5, a switch is made to the TOXLINE file and the generic name for the pesticide is entered. In this case, 160 records are retrieved in which the term Leptophos occurs in the title, abstract or keyword fields.

In Search Statement 2, the CAS Registry Number and a few of the synonymous terms from CHEMLINE are ORed (combined) as search parameters. This results in the retrieval of 257 records containing the desired chemical substance. Thus, the CHEMLINE information effectively increases the TOXLINE retrieval by 97 records. Search Statement 3 is designed to select those citations published in 1977 in which both Leptophos and some variation of the term nerve (nerve, nerves, nervous, etc.) appear. The title and source of one of the five citations meeting these criteria are then PRINTed. Naturally, Statement 3 is not intended to provide a comprehensive search for the biological concept since other terms such as CNS or neurotoxicity are not included. However, the listing of the output from this search could assist the user in selecting biomedical terms relevant to the scope of the search.

CHEMLINE Substructure Searches

The following sections are provided to demonstrate several approaches to substructure searching in CHEMLINE (3-6). Figure 6 shows the structural diagram of Leptophos and the characteristics of a substructure search query based on the chemical structure of this insecticide.

A logical approach to this query would be to search CHEMLINE for the given common name, PRINT the systematic chemical name and select name fragments that correspond to the desired substructure. This is illustrated in the first portion of Figure 7.

The name fragment PHOSPHONOTHIOIC is selected since it is the fundamental or one of the most significant features of the chemical substance. PHENYL, as a name fragment, would explicitly exclude any substitutions on the ring (such as chlorophenyl). This is a requirment for Ring A as shown in Figure 6. The coordination of the two name fragments results in the retrieval of 34 records.

The formula fragments CL and BR along with the name fragment 4 requires that at least one chlorine and one bromine atom be present and the 4-position of a ring (or chain) be substituted. These requirements retrieve 294 records from CHEMLINE.

The intersection of those records retrieved in Search Statements 2 and 3 should satisfy the criteria established for the substructure search (6 records).

```
SS 2 /C?
USER:
"FILE TOXLINE

PROG:
YOU ARE NOW CONNECTED TO THE TOXLINE FILE.

SS 1 /C?
USER:
LEPTOPHOS

PROG:
SS (1) PSTG (160)

SS 2 /C?
USER:
21609-90-5 OR PHOSVEL OR FOSVEL OR LEPTOPHOS

PROG:
SS (2) PSTG (257)

SS 3 /C?
USER:
2 AND ALL NERV: AND 77 (YP)

PROG:
SS (4) PSTG (5)

USER:
"PRT 1 TI, SO

PROG:
1
TI  - NEUROTOXICITY OF ORGANOPHOSPHORUS INSECTICIDES LEPTOPHOS
      AND EPN.
SO  - J ENVIRON SCI HEALTH (B); VOL 12, ISS 4, 1977, P269-87
```

Figure 5.   TOXLINE search

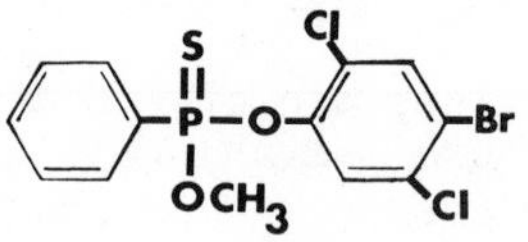

LEPTOPHOS

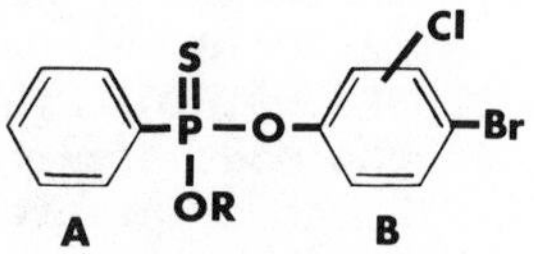

SUBSTRUCTURE

CHARACTERISTICS:

1. NO SUBSTITUTION ON RING A

2. ANY ESTER

3. ONE OR MORE CHLORINE ATOMS ANYWHERE ON RING B IN
   COMBINATION WITH A BROMINE ATOM ON THE 4-POSITION

*Figure 6.   Characteristics of leptophos-related
             substructure search*

```
                        USER:
                        (SY) LEPTOPHOS

                        PROG
                        SS (1) PSTG (1)

                        SS 2 /C?
                        USER:
                        "PRT N1

                        PROG:

                        1
                        N1 - PHOSPHONOTHIOIC ACID, PHENYL-, O-(4-BROMO-2,5-DICHLOROPHENYL)
                             O-METHYL ESTER (9CI)

                        SS 2 /C?
                        USER:
                        (NF) PHOSPHONOTHIOIC AND PHENYL

                        PROG:
                        SS (2) PSTG (34)

                        SS 3 /C?
                        USER:
                        (FF) CL AND BR AND 4 (NF)

                        PROG:
                        SS (3) PSTG (294)

                        SS 4 /C?
                        USER:
                        2 AND 3

                        PROG:
                        SS (4) PSTG (6)

                        SS 5 /C?
                        USER:
                        "PRT N1
```

Figure 7.   CHEMLINE substructure search

ELEMENT NAME	SEARCH ABBREV.	PRINT ABBREV.
SOURCE IDENTIFICATION	(SI)	SI
CHEMICAL SUBST. PRIME NAME	(N1)	N1
CHEMICAL NAME FRAGMENTS	(NF)	--
CHEMICAL DEFINITION	*	CD
CAS REGISTRY NUMBER	(RN)	RN
MOLECULAR FORMULA	(MF)	MF
MOLECULAR FORMULA FRAGMENTS	(FF)	--
MOLECULAR WEIGHT	--	MW
WISWESSER LINE NOTATION	(WL)	WL
SYNONYMS	(SY)	SY
TOXICITY DATA INDEX STRINGS	(IX)	TDKW
TOXIC DATA SOURCE	**	SO
AQUATIC TOXICITY RATING	*	AQ
TOXICOLOGY & CANCER REVIEW	*	TC
STANDARDS & REGULATIONS	*	SR
NIOSH CRITERIA DOCUMENTS	*	NC
STATUS	*	ST
TEXT WORDS	(TW)	--
CITATIONS TERMS	(CT)	SO

*   SEARCHABLE AS TEXT WORDS

** SEARCHABLE AS CITATIONS TERMS

Table I.   RTECS—Unit Record Description

The systematic chemical names are then PRINTed to compare the search results with the desired substructure query. The structural diagrams for the substances retrieved are illustrated in Figures 8 and 9.

CHEMLINE-RTECS Substructure Search

The next example of a CHEMLINE substructure search is for chlorinated dibenzodioxins. As seen in Figure 10, this substructure query is derived from the herbicide contaminant Dioxin. The search can be approached without using any chemical nomenclature through use of the Formula Fragment (FF) and ring information fields (Ring Elemental (RE) analysis and Component Line (CL) formula). The search retrieves forty records; a few of the chemical names are PRINTed to review the output.

Since systematic chemical name fragments and molecular formula fragments are also found in the RTECS file, one can use CHEMLINE nomenclatural output to perform a substructure search that is correlated with a biological concept in RTECS. This would obviate the need to carry the chemical identifiers for the forty CHEMLINE records into the RTECS file.

After accessing the RTECS file (Figure 11), the name fragments common to each name listed in the CHEMLINE search (DIBENZO and DIOXIN) and the formula fragment CL are used to retrieve six records. The RTECS Toxicity Data fields (IX) are then searched for the terms ORAL, RAT and TERATOGENIC and the records containing these terms in the index string are intersected with Search Statement 1 to retrieve two records (Search Statement 2).

Finally, part of the first RTECS record is listed to obtain the toxic data for the chlorinated dibenzodioxin.

Summary

This paper has described several approaches to the utilization of an on-line chemical dictionary file in linking chemical structures to biological information. Through use of nomenclatural, molecular formula and ring screens, the capability exists in CHEMLINE to effectively identify chemical substances with specific structural characteristics. Since CHEMLINE is an integral part of the National Library of Medicine's On-Line Services, it is then possible to correlate the substructural data with biological information and data contained in bibliographic and/or data files existing within the same computer environment.

1
N1 – PHOSPHONOTHIOIC ACID, PHENYL-, O-(4-BROMO-2,5-DICHLOROPHENYL)
ESTER, POTASIUM SALT (9CI)

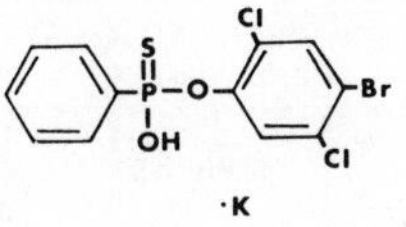

2
N1 – PHOSPHONOTHIOIC ACID, PHENYL-, O-(4-BROMO-2,5-DICHLOROPHENYL)
O-METHYL ESTER, MIXT, WITH
1,1'-(2,2,2-TRICHLOROETHYLIDENE)BIS(4-CHLOROBENZENE) (9CI)

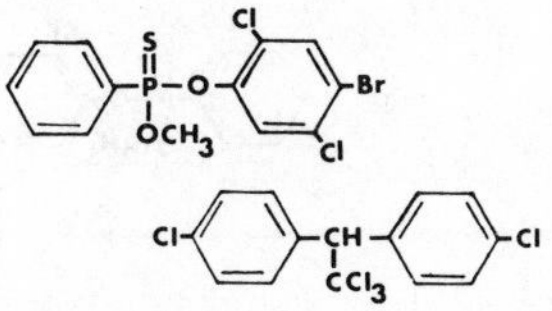

3
N1 – PHOSPHONOTHIOIC ACID, PHENYL-, O-(4-BROMO-2-CHLOROPHENYL)
O-METHYL ESTER (9CI)

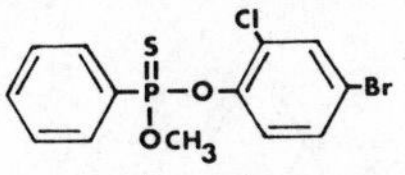

Figure 8. Substructure search results

4
N1 - Phosphonothioic acid, phenyl-, O-(4-bromo-2,5-dichlorophenyl)
  O-methyl ester (9CI)

LEPTOPHOS

5
N1 - Phosphonothioic acid, phenyl-, O-(4-bromo-2,5-dichlorophenyl)
  O-ethyl ester (9CI)

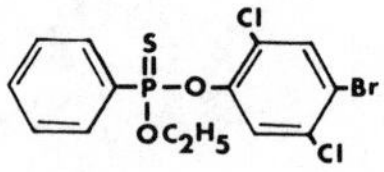

6
N1 - Phosphonothioic acid, phenyl-, O-(4-bromo-2,6-dichlorophenyl)
  O-methyl ester (9CI)

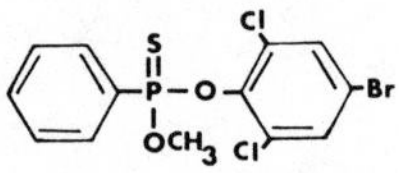

Figure 9. Substructure search results

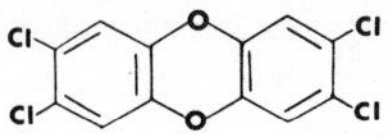

DIOXIN

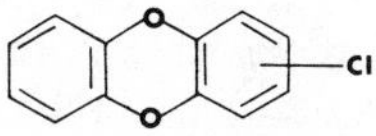

CHLORINATED DIBENZODIOXINS

```
USER:
(RE) C4O2-C6-C6 AND OC2OC2 (CL) AND CL (FF)
PROG:
SS (S) PSTG (40)

SS 2 /C?
USER:
"PRT 4 N1
PROG:
1
N1 - DIBENZO(B,E)(1,4)DIOXIN, TETRACHLORO- (9CI)

2
N1 - DIBENZO(B,E)(1,4)DIOXIN, 1,2,3,7,8-PENTACHLORO- (9CI)

3
N1 - DIBENZO(B,E)(1,4)DIOXIN, 1,2,3,4,6,7,8-HEPTACHLORO- (9CI)

4
N1 - DIBENZO(B,E)(1,4)DIOXIN, HEXACHLORO- (9CI)
```

Figure 10.   Chlorinated dibenzodioxins search

```
SS 2 /C?
USER:
"FILE RTECS
PROG:
YOU ARE NOW CONNECTED TO THE RTECS FILE.

SS 1 /C?
USER:
(NF) DIBENZO AND DIOXIN AND CL (FF)
PROG:
SS (1) PSTG (6)

SS 2 /C?
USER:
(IX) ORAL AND RAT AND TERATOGENIC AND 1
PROG:
SS (2) PSTG (2)

SS 3 /C?
USER:
"PRT 1 TOXDATA
PROG:
SI   - NIOSH/HP32000
N1   - DIBENZO-P-DIOXIN, HEXACHLORO-
RN   - 34465-46-8
SO   - ADCSAJ ADVANCES IN CHEMISTRY SERIES. 120,55,73
TDKW- ORAL;RAT;RODENTS;LDLo;100 MG/KG
      ORAL;RAT;RODENTS;TDLo;100 UG/KG/(6-15D PREG) ;TOXIC EFFECTS
      TERATOGENIC
```

Figure 11.   RTECS search

ABSTRACT

The National Library of Medicine's (NLM) on-line chemical dictionary file (CHEMLINE) is primarily used to enhance the retrieval of bibliographic information associated with chemical substances. This discussion demonstrates the utility of CHEMLINE as a mechanism to link chemical substructures to biological data. Search techniques are developed to integrate classes of structurally related chemicals with toxicity data and information contained in on-line retrieval files such as the Registry of Toxic Effects of Chemical Substances (RTECS) and TOXLINE.

Acknowledgements

The authors gratefully acknowledge the assistance provided by Dr. Henry M. Kissman and Mrs. Joan H. Cepko of the Toxicology Information Program.

Literature Cited

1. Schultheisz, R. J., Kannan, K. L. and Walker, D. F., "Design and Implementation of an On-line Chemical Dictionary (CHEMLINE)", J. Am. Soc. Inf. Sci., Accepted for publication in Vol. 29 (1978).
2. For sale by the Superintendent of Documents, U.S. Government Printing Office, Washington, D.C. 20402 GPO Stock No. 017-033-00271-1.
3. Fisanick, W., Mitchell, L. D., Scott, J.A., and Vanderstouw, G. G., "Substructure Searching of Computer-Readable Chemical Abstracts Service Ninth Collective Index Nomenclature Files", J. Chem. Inf. Comput. Sci., 15 (2) 73-84 (1975).
4. Dunn, R. G., Fisanick, W., and Zamora, A., "A Chemical Substructure Search System Based on Chemical Abstracts Index Nomenclature", J. Chem. Inf. Computer. Sci., 17 (4), 212-219 (1977).
5. "Substructure Searching of Computer-Readable CAS 9CI Chemical Nomenclature Files (Based on Nomenclature in the Ninth Collective Index of Chemical Abstracts) (1972-1976)", Chemical Abstracts Service, Columbus, Ohio, Aug. 1974, 128 pp., ISBN 8412-0204-4, LCN 74-14778.
6. Vasta, B.M. and Spann, M. L., "Chemical Searching Capabilities of CHEMLINE", presented at the 172nd National Meeting of the American Chemical Society, August, 1976.

RECEIVED August 29, 1978.

Chemical and Biological Data—an Integrated On-Line Approach

E. E. TOWNSLEY and W. A. WARR

Data Services Group, ICI Pharmaceuticals Division, PO Box 25, Alderley Park (Mereside), Macclesfield, Cheshire, SK10 4TG, England

For some years CROSSBOW technology (1-5) has been used by ICI Pharmaceuticals Division to search an on-line chemical database. A computerised system for biological control and for the storage and retrieval of biological data has also been in operation for many years. However, the biological database was held separately from the chemical database and there were some interface problems. The advantage of the new integrated chemical and biological system over the previous two separate systems is the ease of access (including on-line access) to all information, both biological and chemical. In the past it was not quite so easy, for example, to follow a biological search with a search of the chemical database, or vice versa.

When the Division installed a Burroughs 6700 computer it was necessary to redesign the databases and convert all programs which had earlier run on a Burroughs 4700 machine. It was decided that development, improvement and conversion should be undertaken at the same time, and an integrated on-line database is now in use. This ever-expanding database at present holds 190,000 chemical compounds from five ICI divisions: data on 267 distinct biological tests: and over 1,000,000 biological test results from Pharmaceuticals Division. Linked to the computer are three Burroughs terminal computers and eight visual display units. The system's uses are chemical and biological data registration and retrieval, and biological control.

The biological control system manages the processing of samples submitted for biological assay. Data Services Group takes a compound from the chemist, weighs appropriate samples for all the necessary biological tests and collects test results from the biologists.

The first step in the biological control sequence is to register the structure of a compound, and if it is novel, to allocate it the next available six-figure M number, say, for example, M100,000. This first sample is referred to as M100000/01 where 01 is called a stroke number. If the compound is not

0-8412-0465-9/78/47-084-073$05.00
Published 1978 American Chemical Society

novel, the sample will be allocated the next available stroke number to the previously allocated M number, e.g. M80996/09.

The next step is to register the property data (see Figure 1) of the given sample and to allocate a sample storage number, which is the shelf storage number for the stock bottle.

The handling codes recorded on the property file are two-letter codes indicating whether any unusual handling precautions need to be observed (e.g. TX for "toxic", KC for "keep cold").

Intention-to-test data is held on a dynamic file whose record layout is shown in Figure 2. For each sample/test combination an intention-to-test record is created and used to manage the progress of the sample from the chemist's initial submission through to the collection of test results from the biologist, after which the intention-to-test record is removed.

Most of the data in Figure 2 is self-explanatory. A test is represented by two or three characters (e.g. AB or AB3 for antibacterial tests) for external use and a number for machine use.

The sample property record must be created before an intention-to-test record can be made because data needed for the latter record is abstracted from the former (compare Figures 1 and 2).

The allocation of a priority is important because it ensures that compounds are progressed through the system in such a way that important project compounds are always selected in preference to non-project compounds where there are more test submissions than a biologist can cope with at any one time. Four priorities are used. Compounds made by the chemist for testing in connection with his own project are given priority 1 (the highest). The chemist may elect to send the same compound for tests unconnected with his project and the intention-to-test records for these will have priority 2. Compounds the chemist has selected by means of a substructure search around a "lead" compound, will be tested on priority 3 and screening compounds randomly selected by Data Services Group are tested under priority 4 (the lowest). The date of submission is recorded so that, for a given priority, records created earlier are processed before those created later.

The status of a sample is described in Figure 3. When an intention-to-test record is created it has the status value set to 1. This value is updated stepwise from 1 to 6 as the sample progresses through the weighing and documentation process, followed by submission to the biologist and finally reaches the result collection and notification stage. When the test control function for the sample is complete, the intention record is removed from the file.

M number/stroke number   Key: Index sequential
Company registry number
Suffix pointer (e.g. 0201 or 0001 in Figure 7)
Sample available/not available/reserved
Sample storage number
Chemist's initials and notebook reference
Salt data
Solubility codes
Handling codes

Figure 1.   Property file (standard data set)

M number
Stroke number   } Key 1: Index sequential
Test number
Priority   } Key 2: Index sequential
Date
Status
Test letters
Sample storage number
Chemist's initials and section
Handling codes

Figure 2.   Intention-to-test file (standard data set)

Status	Meaning
1	Sample for this test unweighed and undocumented.
2	Sample for this test weighed but undocumented.
3	Sample for this test weighed and documented.
4	Sample for this test submitted to biologist.
5	Test result has been received from biologist.
6	Test result notified to section leader and intention-to-test record ready for removal.

Figure 3.   Status of sample on intention-to-test file

From the intention-to-test file is generated a matrix
report which details the current state of sample progression
against test. It reports the numbers of samples for each test
at each management level, up to and including status four,
subdivided by priority. Enquiry routines give more detailed
information at the specific sample level.

Biological tests are of two types - current and screen.

A current test is one which requires project compounds to
pass as quickly as possible from chemist to biologist, and such
compounds must be weighed and documented immediately after
input. Usually there is only a small number of samples to be
processed each week for one of these tests.

A screen test is one which requires a regular supply of a
fixed number of samples (e.g. 200 per week). This quantity is
made up of chemist's project compounds and other currently
submitted compounds, and the balance is provided from priority
3 and 4 submissions.

To cater for the two types of test there are two ways in
which samples are selected for progress to weighed and documented status and then to submission to the biologist.

The first produces weighing instructions, bottle labels
and test documentation for samples selected by M number -
essentially current project compounds. Usually, all the test
samples from one stock bottle are weighed at the same time,
irrespective of test type, so as to minimise bottle handling.
However, documentation for screen tests is produced at a
later stage. Thus the status of current test records is
updated to 3 and that of screen test records to 2, by this
"weighing routine".

The second weighing routine, which is geared to screen
test requirements, selects status 1 and 2 records by test. For
each test the samples are selected in priority order, and
within priority, by date. Again, weighing instructions and
bottle labels (where needed) and test documentation are
produced together with two different working documents which
detail sample numbers and related data, one for the biologist's
use and one to be used within Data Services Group.

A test details file is maintained which holds constant
data pertaining to each test, as shown in Figure 4. When test
result data is returned it is validated against the details held
on this file.

At the same time as an intention-to-test record is created,
a "result" entry is created on the summary result file (see
Figure 5) with the result area left blank. As the sample
progresses to submitted status the blanks are replaced by "S".
Finally this area will be occupied by the biologist's summary

test results (e.g. A for active or I for interesting) for up to four subdivisions of his test.

A screen file (Figure 6) is maintained at the compound level as opposed to the sample level. Each compound is allocated a sample availability bit and one bit per test which is used to indicate whether or not any sample of the compound has been submitted for that test. This file is maintained because the summary test result file is very large and before it is searched for test result information the screen file is quickly scanned so as to reduce the number of accesses that need be made to the larger file.

To return to the beginning of the registration process, a chemical structure is input as a molecular formula and a Wiswesser Line Notation (WLN) (6-7). The first on-line routine generates a molecular formula from the input WLN, and if this agrees with the input molecular formula a novelty check is carried out against the WLNs already on file. If the compound is novel its WLN is registered and a fragment screen is generated from the notation and stored (separate from the WLN) on a direct data set with the company registry number (CR number) as key. There are 148 designated fragments, the presence of which are indicated by bits set to one rather than zero. Each fragment is related to a single WLN character, or a group of them, or to a feature of a WLN cited ring. Ring substituent fragments are differentiated from similar fragments in a chain.

The way WLN data is held is shown in Figure 7. It can be seen from this that a CR number represents a WLN, or a two-dimensional structure whereas a divisional reference number (e.g. an M number) represents a compound. Sometimes information such as stereochemistry, has to be suffixed to the WLN proper to give a full representation of a compound. In the many cases where there is no suffix, CR number and divisional reference number are equivalent.

The combination of WLN-last-16-inverted plus WLN length was chosen as key because for a very high proportion of the records on the file this key is unique. The mean length of notations registered on the database is 21 characters. The main WLN file holds up to 34 characters of notation. A small number of WLNs (about 7%) have to overflow onto a subsidiary file. Thirty-four characters of WLN are held on the main file, because that number reduces wasted space on the main and over-flow files to a minimum.

The database is designed so that it may be accessed by WLN, divisional reference number, CR number or molecular formula.

```
Test number        Key 1 - direct
Test letters       Key 2 - index sequential
Flag - is test current or screen?
Biologist's initials
Weight of sample required for test
Stationery requirements
Test result format
```

Figure 4. Test details file (direct data set)

```
M number/stroke number/record number     Key - index sequential
Test letters - priority  ⎫
               date      ⎬  occurs once for each
               chemist   ⎪  test on this sample
               summary result ⎭
```

Figure 5. Summary result file (standard data set)

```
M number        Key - direct
Flag - sample available/not available
Tested/untested bits for 480 possible tests
```

Figure 6. Screen file (direct data set)

There is a separate data set for compounds which cannot be coded into WLN (e.g. seed extracts and reaction products of unknown structure) and these are stored on a random data set with divisional reference number as key.

The CROSSBOW chemical search system (1-5) is a multilevel one. Search of the bit screens is first carried out and this quickly and cheaply reduces the file to ten per cent (or less) of its original size. There will almost certainly be many false drops but most of these can usually be removed by string search of the WLNs and/or molecular formulae and/or reference numbers. String searching is slower and more expensive than bit search. Connection table generation and atom-by-atom search of the connection tables (the third search level) are still slower and even more expensive, but the atom-by-atom search program is a very powerful tool which is used in about 80% of all searches. The CROSSBOW connection tables for the hits from any search are finally used as input to a structure display program.

Bit and string searching is an interactive process. The paramaters are input on a VDU, the hit count is displayed on-line and, if necessary, the search parameters can be modified and the search repeated. At the end of a session, search hit files are merged as required and connection table generation, atom-by-atom search and structure display are run batchwise.

An average bit and string search takes about five minutes (searching 190,000 compounds). The CROSSBOW connection table generation program handles 1500-2000 compounds a minute, and the atom-by-atom program searches 600 compounds a minute. Well over 90% of notations are amenable to connection table generation and over 90% of the compounds on the database can be structurally displayed directly from the connection table. (The remainder are held on the database in a file of "difficult" structures.)

A bit search is specified using combinations of AND, OR and NOT logic, and also nested AND-within-OR logic, as required. String search allows the same logical operators, but in addition there are ignore and followed-by facilities and other specialised syntax to simplify the specification of several alternative strings.

Figure 8 shows a typical bit and string search. In the bits search, fragments 50 and 137 must both be present. Neither fragment 134 nor fragment 135 may be present, and nor may be fragment 136. The OR clause states that if the combination of (either 17 or 18) and (either 71 or 72) is not present, then, alternatively the combination of (either 67 or 68) and (either 28 or 29) must be present. The Wiswesser strings which could represent this substructure are:

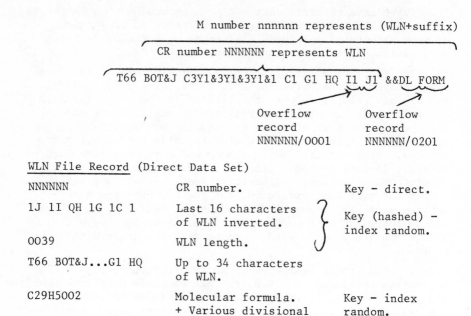

Figure 7.   *Storage of chemical structural information*

X = OCH_3 or NHR (R≠H)

Only the one ring present.

Subst halogen	50		
One benzene ring	137		
Subst CO.O	71	72	
Subst CO	67	68	
Chain methyl	17	18	
Chain NH	28	29	
Ring systems	134	135	136

Parameters

AND,BITS "50 137". NØT, /134 135 136/.
ØR,/71 72/ /17 18/,/67 68/ /28 29/.
ØR,WLN -1/1-"1OVR","MVR"," "#%##V##OM#.END

Figure 8.   *Bit and string search*

Notations beginning "1OVR"
or notations containing "MVR"
or notations containing " xVO" or " xVM"

where x = any character but space (represented by % in
Figure 8).

The last two strings are combined together, the hash
marks enclosing characters which are alternatives.

Atom-by-atom search is resorted to if the string search
of WLNs and/or molecular formulae and/or reference numbers is
not specific enough. This will occur if, for example, the
substructure is very branched or may be partially or wholly
embedded in a ring. An atom-by-atom search may also be
needed if there is a specific positional relationship between
two groups and that relationship cannot be accurately defined
by WLN sub-strings. The structure in Figure 8 falls into the
last category.

An atom-by-atom search to fix the spatial relationship
between the halogen and the carbonyl groups is shown in
Figure 9. E, F, G, I, D, T, U, O, Q and M are the possible
ways the nodes 1-7 could be represented in the CROSSBOW
connection table, a node being an atom and its surrounding
bonds. NOTR stands for "not in a ring" and RINGSA for a
single benzene ring (S) of the type shown in the RINGSCREENA
parameter. The RINGSCREENA parameter designates a benzene ring
which has substituents at the 1 and 3 positions and optionally
substituents at other positions also. The seventh node must
not be terminal (or the Q would represent OH or the M, NH_2) so
the L before the QM indicates that this node must be linking.
The network parameter B65X, indicates that the seventh node is
attached to the fifth.

If desired, a biological search may be carried out before
or after the chemical search. This may perform the function of
reducing the total search output, or it may be carried out
merely to produce all the test and sample data for each
compound which is a hit on the chemical search.

The operator MNOS is used to specify the range of numbers
to be searched and, if required, to limit the number of hits.
The operator OUT is used to specify what sample and result data
is to be printed for each hit. The operators EVER and NEVER
(similar to OR and NOT) are used to search the data on the
property file and the summary result file, using the screen
file, if relevant, as a preliminary screen. Thus, for example,
one may search for all the active anthelmintics Dr Robinson
made and reserved in 1977; or all the solids in large size
bottles which are active in an anthelmintic test but have not
been tested in any of the liver fluke tests.

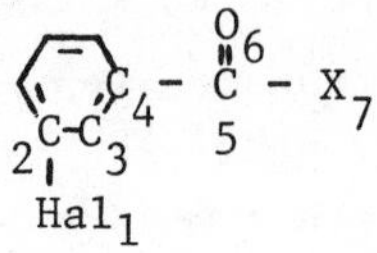

$$X = NH \text{ or } O.$$

Parameters

```
RINGSCREENASTCTCCC
1NOTR    EFGI
2RINGSA  T
3RINGSA  DT
4RINGSA  T
5NOTR    TU
6NOTR    O
7NOTR    LQM
B65X
```

Figure 9.   Atom-by-atom search

Figure 10.   Biological search input

```
AA MNOS,1-20000=1000.
   EVER,(ACTIVE"AB2")(AVAIL).
   NEVER,(TESTED"AB3").
   OUT,BSLONG BRLONG ALL.END
```

AA 13/03/78 *123456   SEARCH OUTPUT   M131833
MOLECULAR FORMULA
C9H10O3
WISWESSER NOTATION
QYR&VO1

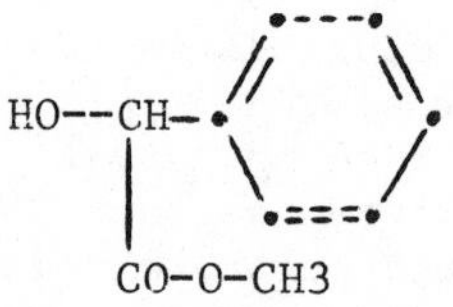

L FORM

OTHER REF NUMBERS
R12345

STROKE NUMBER	TEST LETTERS	RESULT	WEEK	YEAR	CHEM.	SECT.	PRIORITY
01	LF1	N	48	77	WAW	99	2

Figure 11.   Search output

Figure 10 shows a biological search for 1000 compounds in the range M1 to M20000, these compounds being active in the AB2 test, having samples available and having never been tested in the AB3 test. The OUT parameter states that full sample data must be output and full result details for every test to which each compound has been submitted.

The output from a chemical search is 8" x 5" cards, each card bearing a structure drawn from the connection table by the computer. An example (not related to the two searches above) is shown in Figure 11.

Planned Developments

Wiswesser Line Notation is likely to be used at ICI Pharmaceuticals Division for many years to come, not least because of its use in commercial databases bought by ICI, for example the Index Chemicus Registry System tapes (8).

We are unlikely to put this vast database on-line, although we regularly manipulate it and search it using the CROSSBOW system. We do, however, have immediate plans to augment our on-line database by adding various sample property and test result files for two more ICI divisions, and by adding the Commercially Available Organic Chemicals Index (CAOCI) being built up by a number of European companies.

Acknowledgement

We would like to express our thanks to Paul Bowler (who was instrumental in the design of the database) for his helpful reading and criticism of this paper.

Literature Cited

1.   Hyde, E., Matthews, F.W., Thomson, L.H. and Wiswesser, W.J. - "Conversion of Wiswesser Notation to a Connectivity Matrix for Organic Compounds". J.Chem.Doc., V.7(4), p.200-204 (1967).

2.   Thomson, L.H., Hyde, E. and Matthews, F.W. - "Organic Search and Display Using a Connectivity Matrix Derived from the Wiswesser Notation". J.Chem.Doc., V.7(4), p.204-207 (1967).

3.   Hyde, E. and Thomson, L.H. - "Structure Display". J.Chem.Doc., V.8, p.138-146, 1968.

4.   Eakin, D.R. - "The ICI CROSSBOW System" and Ash, J.E. - "Connection Tables and Their Role in a System" in Ash, J.E. and Hyde, E. - "Chemical Information Systems". Horwood, 1975.

5.   Eakin, D.R., Hyde, E. and Palmer, G. - "The Use of
Computers with Chemical Structural Information: ICI
CROSSBOW System". Pesticide Sci. p.319-326, 1973.

6.   Smith, E.G. and Baker, P.A. - "The Wiswesser Line-Formula
Chemical Notation (WLN)". 3rd edition, CIMI, New Jersey,
1975.

7.   Baker, P.A., Nichols, P.W.L. and Palmer, G. - "The
Wiswesser Line-Formula Notation" in "Chemical Information
Systems" (loc.cit. at 4).

8.   ICRS Ⓡ tapes contain data from Current Abstracts of
Chemistry and Index Chemicus^{TM} and both are available
from the Institute of Scientific Information, 325
Chestnut Street, Philadelphia, Pa. 19106, USA.

RECEIVED August 29, 1978.

Use of Proprietary Biological and Chemical Data at Merck & Co., Inc.

I. M. R. EGGERS, W. B. GALL, F. A. CUTLER, JR., and H. D. BROWN

Merck Sharp & Dohme Research Laboratories, Rahway, NJ 07065

Like other pharmaceutical laboratories, Merck Sharp and Dohme Research Laboratories is amassing an enormous amount of proprietary biological and chemical data. Moreover, the scientific staff at MSDRL is using these data in increasingly sophisticated ways in their search for products to improve human and animal health.

An outline of the systems (shown in Figure 1) will be helpful for understanding this usage.

Biological Data

Biodata Systems Structure. The beginning of the biological data system grew out of the notion that it would be helpful to have all the data on a given compound immediately available in a single listing. From the original hand-posted card file, a punched card record was developed, using the concept of one record for one test. In 1968 a computer tape record was started. This computer-stored file now consists of 2.5 million records in several hundred tests representing 135,000 compounds. The basic record as it appears for entry into computer storage is outlined in Figure 2.

Each record answers the following questions, either by summarizing the experiment itself or by linking or pointing to other files.

1. When was the test done?

2. What was being tested? The 11-digit L-number is the link to the Chemdata files which will supply the chemical name and, if desired, the structure.

3. How was the test performed? The 4-digit test number provides the link to the file of descriptions of the several hundred tests currently in computer storage. Variations on the basic test, such as different routes of administration

0-8412-0465-9/78/47-084-085$05.50
Published 1978 American Chemical Society

Biological Data
    Biodata Systems Structure
    Records Acquisition
    Search Strategies and Examples
Chemical Data
    Chemical Structure Information Systems (CSIS)
    Chemical Names
    Compound Repository
    Transaction File
    Chemdata Enhancements
    Transaction File/Inventory
Interaction

Figure 1. Outline

Cols.			
1-6	L-Number	Must be present and numeric
7-8	Salt Form	Must be present and numeric
9	Check Letter	Must pass program which checks cols. 1-8
10-11	Batch	Must be numeric, if present
12-15	Test Number	Must be present and numeric
16-40	Test Name	Free Form
41-43	Protocol Code	Must be present and numeric
44-72	Test Result	Free Form
73	Response Code	Must be present and numeric
74	Dose Code	Must be numeric, if present
75-80	Test Date	Must be present - month, day, year

Figure 2. Basic record—biological data

of test compound, or dosing schedules, are defined by the 3-digit protocol codes. At present, the number of variations is in excess of 2,400. Definitions are on each Data Sheet, <u>very</u> abbreviated.

4. <u>Why</u> do we run this test? The rationale is a part of the protocol, which is written in detail, dated, and signed by the biologist in charge of the test, and is filed in the Biodata Department.

5. <u>Where</u> was the test done and <u>who</u> did it? These items are noted on the Data Sheet, as well as being incorporated in the Protocol write-up.

There are 53 columns called "free form", whose content varies with the test at hand. The specific format of each test is described on the Biological Data Sheet (see Figure 3).

When properly filled out, these forms constitute the set of directions for the keypunch operators. These are made out by the Biodata Department in cooperation with the biologist responsible for the test and the chemist responsible for supplying materials to the test.

These "free form" items make the system basically open ended. <u>Among</u> the parameters entered into these "free form fields" are those listed in Figure 4. The dose units may be explicit, abbreviated, or implied.

Occasionally, the number of parameters to be recorded is so large that the actual record is simply a series of numbers. If necessary, two or more records are generated.

All of these records constitute <u>data</u>, as opposed to <u>documents</u>, and this difference is most significant. The Protocol file is a document file. In the planning stage is a Term Index to the protocol file, containing in a searchable form such items as species, route of administration, whether enzyme, virus, yeast, or other type of test, date of first record in computer storage, name of investigator, dose units, and possibly others.

Various other files are maintained for such purposes as housekeeping, analysis of activities, and cross reference (see Figure 5). Also, there are some manually-posted files of biological data which predate the computer system.

That is the structure of the system. Data must be acquired to build it in the first place and then to maintain it on a current basis.

<u>Records Acquisition</u>. Records may be sent to Biodata from any Merck laboratory site or outside investigator. These may be in the form of: (1) copies of laboratory notebook records (the most common), (2) punched cards from systems not compatible with

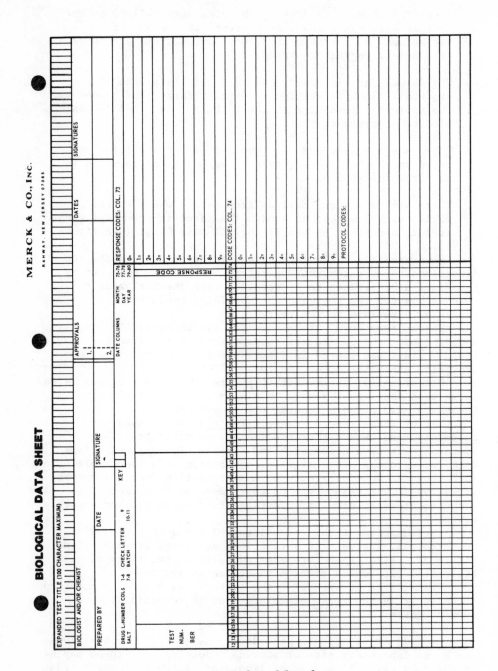

Figure 3. *Biological data sheet*

ours, (3) documents already distributed to the appropriate members of the scientific staff, (4) magnetic tapes, or (5) manual card files. We generally accept biological data in whatever form is most convenient and ask that no one do any extra work beyond making copies.

Approximately 96 percent of the keypunching for entry to the computer is done within the Biodata Department. Documents are usually not supplied with card column headings. The data entry personnel are expected to use a fair degree of judgment in reading documents from which data are to be keyed. They are able to convert from one dosage unit to another, to calculate percentages based on control values, to recognize various types of errors and omissions, and to use many other skills obviously well above the level of the average keypunch operator. They also type, take dictation, operate electronic data processing equipment, use microforms equipment and the computer terminal and sometimes write computer search programs of fairly elementary nature.

Some data arrive at the department already on punched cards or a magnetic tape. In this case, the conversion to the described format, plus the transfer into the master computer file is handled by members of the Biodata staff.

Each computer-stored record has links to the input which produced it. The date and test number are the primary links, but, to the extent possible, experiment number and/or notebook and page number are also recorded. The test number plus the protocol code serve to reference the file of protocols. An elaborate audit trail is kept for changes which need to be made in records. All accessory files are created within the department.

Data are input into computer storage on a daily basis in batch mode, and all data are accessible as soon as entered. This daily input creates a monthly file, which seldom runs over 30,000 records. At the end of each month, it is merged into the total file.

The file has been defined and the data records have been acquired. The next section deals with how they are used.

Search Strategies and Examples. We have the capability of (a) selecting a group of data records such as those keyed by the same compound number, test number, date, combinations thereof, or other conditions, (b) comparing those records according to some defined constraints and logic, and (c) using the resulting analysis to retrieve for printing the appropriate records from the same or other files. In complex situations, previously written programs assist the process.

The queries directed to Biodata fall fairly neatly into eight categories (see Figure 6). Usually some interface with one or more of the Chemdata files is necessary. The connecting link is the identification number of the material, the L-number.

It is important that the question be phrased so that it is completely unambiguous. We like to negotiate each search individually with the requesting scientist. It is important to the progress of the research that he gets what he needs, not necessarily what he started out asking for.

There are operational programs which are often used for searches of types 2, 3, 4 and 6 (see Figure 6). No changes may be made in output headings or formats and, except for type 6, no response, dose, or protocol conditions may be imposed. Names of compounds are always added, and we _may_ ask that the computer include structural diagrams. If additional sophistication is required, a custom program will be written.

The first four types are probably self-explanatory. The third type in Figure 6 is most heavily used. At the end of each monthly merge, all 2.5 million records are microfilmed, using COM (_C_omputer _O_utput _M_icrofilm). It is thus possible, by using a microfilm reader/printer, to answer this sort of request within a half hour. Suppose the question is "What do we know about a compound numbered L-475,878?" The answer is illustrated in Figure 7. This typifies the answer to the very first thing we wanted to be able to do - to put all the data on a given material together in one listing.

Search type 4 in Figure 6 is ordinarily requested in order to review the performance of a particular set of compounds in certain test systems. It also provides a list of those compounds which were _not_ tested in the defined set. This list can then be sent to Chemdata with a request for samples to be submitted for the test(s) in question, if the asking scientist desires.

Search type 6 in Figure 6 is illustrated by the hypothetical correlation shown in Figure 8. All compounds which have been examined in two groups of tests are compared as to specified responses in each group. Here 3,600 compounds were tested both _in vitro_ and _in vivo_. The data show that 87 percent of the compounds active in the test tube were active in mice and that less than 1 percent of the compounds inactive in the test tube were active in mice. Such correlations are useful in developing cost effective testing programs, as well as establishing unexpected relationships. Most relationships are not so obvious as this, and may require statistical analysis.

Single record selections (Type 5 in Figure 6) imply that there is no need to consult any other record to determine whether you wish to save the one at hand. An interrecord comparison requires, on the other hand, that a set of records is to be retrieved _only_ if two or more records, either in that set or another, meet some defined criteria.

"Additional complications" (Type 8 in Figure 6) covers a multitude of sins. The program may be so horrendous that the computer run must be split up into sections. Many subfiles may need to be built. Interface may be necessary with the output of a structure search from Chemdata.

Infecting organism and strain and animal species
Route of administration of drug and challenge
Injection site
ED_{50}, LD_{50}
Therapeutic Index, MIC
Percent of control
Side effects - three-digit code
Organ weights, body weight
Doses or concentrations in various units, as mg/kg, mcg/ml

Figure 4. Examples of data captured

Test number file (accessible by computer terminal)
Index to Data Sheet file (accessible by computer terminal)
Search statistics file (on punched cards)
Organism cross-reference file (on punched cards)
Cross reference files for:
    L-numbers
    MK-numbers
    NCI-numbers
    any other numbers necessary

Figure 5. Other files

1. No data, e.g., only protocols, counts, etc.   20
2. All data for a previously defined set of tests   55
3. All data for a previously defined set of compounds 134
4. All data for a previously defined set of compounds
   in a previously defined set of tests   121
5. Single record selections, with or without file
   matching   70
6. Correlations   7
7. Interrecord comparisons   185
8. Additional complications   <u>22</u>
   Total   614
   Number requiring structures: 148

Figure 6. Types of searches released—1977

PRINT OF BIOPROFILE ON L-475,878 ON MICROFILM

L-NUMBER	SALT	C L	BATCH	TEST NUMBER	DESCRIPTION	PROT. NO.	RESULT	RD EO	DATE OF TEST MO DA YR	KEY TO TEST CATEGORY	DATE OF ENTRY INTO COMPUTER STORAGE
L-475,878	00	E	02	1003	RNA DEP. DNA POLYM. 72-116	002	SAMP/CONT 0.50 @ 50.0 MCG/ML	13	11 27 72	17	11 28 72
L-475,878	00	E	02	2342	PROTEUS VULGARIS VITRO	002	INACTIVE AT .100 MG/ML R/D	04	08 15 67	15	01 01 70
L-475,878	00	E	03	2725	HYPERTEN. RAT 77-510 (2)	001B	13.0 @ 80. MG/KG	0112	08 14 77	12	09 15 77
L-475,878	00	E	02	2891	RAT DIUR. 872 108	001	0000 0011 C O 05.0-050 MG/KG	0	09 21 66	20	06 12 70

12/22/77
(date of microfilm from which print was made)

Figure 7. Print of bioprofile on L-475,878 on microfilm

Except for operational programs such as the four noted above, all questions involve custom programs or manual searches or possibly a combination of several methods.

Requests to be repeated fairly often will be catalogued in our search library.

Roughly 65-70 percent of searches involve one-time programs. The needs of researchers vary greatly, depending upon subtle differences which cannot be defined adequately until they begin to interact with the data bases. Response to this constantly changing need is of utmost importance in a research environment.

The custom programming is handled by members of the Biodata Department. There are four technical people and five clerical people. All are presently writing computer search programs upon demand. These programs are assigned according to their current degree of expertise.

The 614 searches run by Biodata in 1977 have been broken down into the method used to handle them in Figure 9.

The total, 805, is greater than 614, because some required more than one method. The 111 persons requesting these 614 searches were 46 chemists, 50 biologists, and 15 from legal, management, marketing and other information science areas.

The term "The Data Analyzer" in Figure 9 refers to a very efficient FORTRAN-based software package marketed by Program Products, Inc., Nanuet, New York. It is used for nearly all the custom programs. Operations are mostly in batch mode, although there is a teletype terminal, which is suitable for subsets of the master file and for some smaller, accessory files.

Each of the 80 characters in the record is searchable, as is any combination thereof. Fields may be defined as required by each search. Subfiles of records or portions of records may be created. Computations may be performed on all numeric fields, bearing in mind that a numeric field in one test number may be alpha in another. Comparing, counting, sorting are performed as necessary. The records in computer storage may look very different from those presented to the requesting scientist, depending upon the need of that scientist. The final output is tailored to the needs of the scientist.

The six records in Figure 10 were combined with names and structures from the Chemical Data Files, reformatted, other information added, and a report produced as shown in Figure 11. This is an example of what the scientific staff is presently receiving. The query here was:

Produce a report of last week's rat hypertension data using appropriate headings for each compound in this order (left to right) any combining compound number, rat type, onset time, dose, route of administration, number of rats, codes for blood pressure change, duration of blood pressure change, heart rate change, and date. Omit book and page reference, test number and vehicle code and add an expanded test title as a heading. Define all codes on each page of

Active in Mice and Active in Test Tube 153 Compounds	Active in Mice and Inactive in Test Tube 19 Compounds
Inactive in Mice and Active in Test Tube 27 Compounds	Inactive in Mice and Inactive in Test Tube 3401 Compounds

Of the 180 (153 + 27) compounds active in test tube, 153 (or 87%) are also active in mice.

Of the 3420 (3401 + 19) compounds inactive in test tube, 3401 (or 99%) are also inactive in mice.

Figure 8. Correlation study (hypothetical)

Hand Look Up	103
Personal Contact	4
EDP Machine	46
Terminal	3
Operational Program	172
Microform	179
The Data Analyzer	290
Structure Driven	3
Other	5
Total	805

Figure 9. Method of solution

```
62734600L012725HYPERTEN.RAT    72-15(2) 112        20.   MG/KG5332 121672
62734600L012725HYPERTEN.RAT    72-15( ) 112        20.   MG/KG  92 121672
62734600L012725HYPERTEN.RAT    73-39( ) 112          .08MG/KG0004 011373
62734600L012725HYPERTEN.RAT    73-38(2) 112          .02MG/KG0004 011373
62734600L012725HYPERTEN.RAT    73-38( ) 112 06.0     .02MG/KG0214 011373
62734600L012725HYPERTEN.RAT    73-36( ) 112 01.0     .08MG/KG0314 011373
```

Figure 10. Records in computer storage

the report. Add names and structures. Distribute 19 copies
to the list of persons designated by Dr. Sweet, who will
write a one page summary cover memo. (This particular
report was 90 pages long.)

Another type of question involves no data display at all,
but can involve just as much programming. An example is shown
in Figure 12.

The requestor now can readily see that, of the 223 compounds tested, 9 produced a response code of 3, and 1 produced a
response code of 1. Note that this may be only 9 compounds,
since obviously there was more than 1 test per compound.

Another example might involve a new test in chicks and
there are no guidelines except, perhaps, a tight budget. What
compounds are there for which a physical sample exists (at least
250 mg), which were not toxic to chicks at 0.1% in the diet? The
answer to this question assures that the researcher won't kill
chicks trying to find non-toxic levels of experimental drugs.

Chemical Data

The Chemdata Systems at Merck Sharp & Dohme Research
Laboratories have grown over the past dozen years from a few
manual files to a complete network of interfaceable, searchable
data bases. For the purpose of comparison, the original systems
consisted of: (1) a chemical structure file which utilized the
classification system which was developed by Dr. F. Wiselogle (1)
in 1946; (2) a manual index card file of the sample repository;
(3) transaction summary sheets containing manually posted
transmittals and; (4) a chemical name file which was a portion
of the record of the transaction summary sheets.

Our current computerized individual systems, including the
original design, content, search and interface capabilities with
other systems, are described, followed by the enhancements that
were added as we gained experience from user interaction. An
overview summarizes exactly how the research scientist at MSDRL
utilizes the systems and how he has realized benefits therefrom.

Chemical Structure Information Systems (CSIS). The most
important system is our Chemical Structure Information System.
The details of this system have been published (2). However,
some major highlights deserve review.

In this system both the updating and the querying of the
chemical structure file begin with typing the structures on a
Magnetic Tape/Selectric Typewriter equipped with a typing element bearing appropriate bonding characters. The tapes are
processed by computer in batch mode.

For file updates the structures are subjected to numerous
validity checks and are analyzed to create a connection table
and assign bit screen elements. Before entry into the master
file, newness to the file is established. The entire input

Antihypertensive Evaluation Codes* Rat Assay Report No. 256	Blood Press. Eff. 1 = 20-29 mm Decr 2 = 30-39 mm Decr 3 = GT 40 mm Decr 8 = GT 20 mm Decr 9 = Toxic 0 = No Effect	Blood Press. Duration 1 = Lt 4 hrs 2 = 4-7 hrs 3 = GT 7 hrs	Heart Rate 1 = Quest. Change 2 = 60-100 B/M Incr 3 = GT 100 B/M Incr 5 = GT 100 B/M Decr

L-627,346-00L
$C_{17}H_{22}F_3N_3O_2$ Mol. Wt. 357.37
N S
Baldwin, J. J. M1536-1105-IV

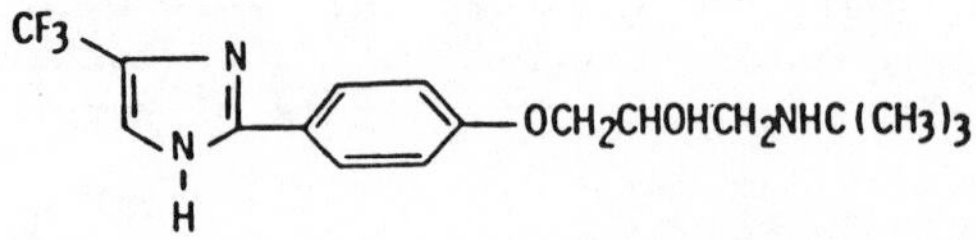

2-[4-(3-tertbutylamino-2-hydroxypropoxy)phenyl]-4-trifluoromethylimidazole

BT	Rat Type	Onset Time Hrs.	Dose Mg/Kg	0 = P.O. 1 = I.P. 2 = I.V.	No. of Rats	Blood Press. Code	Blood Press. Duration Code	Heart Rate Code	Mo	Da	Yr
01	SH		20.	1	2	3	3	5	12	16	72
01	SH		20.	1		9			12	16	72
01	SH		.08	1		0	0	0	01	13	73
01	SH		.02	1	2	0	0	0	01	13	73
01	SH	06.0	.02	1		1	2	0	01	13	73
G1	SH	01.0	.08	1		1	3	0	01	13	73

Figure 11. Final report

408 = Count of tests for the period
223 = Count of compounds

 1 = Count of tests for response code 1
 1 = Count of compounds for response code 1
 0 = Count of tests for response code 2
 0 = Count of compounds for response code 2
15 = Count of tests for response code 3
 9 = Count of compounds for response code 3

Figure 12. Bimonthly report for Test X

record, which includes other descriptive information, is stored in the computer.

Figure 13 represents a chemical structure record as it is input to the system. The L-number in the upper left is the registration number. This is followed by the molecular formula, then any stereo descriptors, reference to source, and finally the chemical structure. This is a typical single record. The file contains approximately 135,000 such records. The year of registration is added to the record automatically at the time of its input to the system.

Tapes bearing substructure queries are processed similarly through edit and automatic bit screen assignment. Although answers may be selected as the result of bit matching alone, the search procedure is normally allowed to proceed past this stage to the atom-by-atom comparison.

Figure 14 lists the major substructure retrieval capabilities.

Figure 15 lists the search parameters which may be invoked in order to provide greater specificity for search queries.

The CSIS system, as illustrated in Figure 16 extracts ancillary data from the chemical name file, the sample repository file and the biodata file. In the case of biodata, it may be specified that "ALL" biodata be printed for the specific compounds retrieved. Additional criteria such as dose and/or activity level requirements may be included. The system is capable of interrogating a customized dictionary of biological test titles singularly or in combination, which when numerically designated, prints only those test results stored within that definition.

This data base is searched nightly in batch mode and is updated on a weekly basis. The structure file is capable of being linked with other MSDRL files by providing structural identification for reports upon request throughout MSDRL.

Chemical Names. Figure 17 illustrates the format of the Chemical Name File.

Each record contains the registry L-number followed by the chemical name which in most cases observes CA guidelines. The length of the name is limited to 180 characters. This file is searchable by L-number and in some cases by text recognition of portions of the name utilizing programs written by the departmental staff.

Compound Repository. Figure 18 represents the record format of a compound repository record.

The registry number is followed by the location descriptors of each specific sample. The file is searchable by L-number and contains 125,000 records sequenced in L-number order.

Transaction File. The most recently developed system is

L-590,226-00A
$C_{19}H_{16}ClNO_4$
NS
WITZEL, B. M5247-118-5

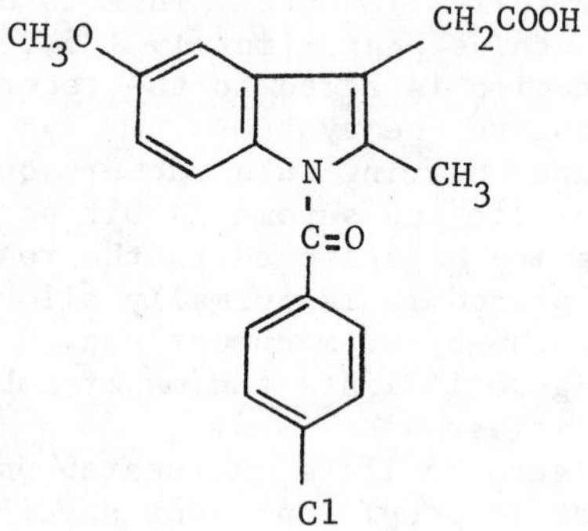

Figure 13.   CSIS input record

Total Structure
Fragment Structure
Multifragment Structure
L-number (Specific or Range)
Year Plus Fragment Structure
Reference Plus Fragment Structure
Exact Molecular Formula Plus Fragment Structure
Molecular Formulae (Minimum or Exact)

Figure 14.   CSIS retrieval capabilities

Statements of indefinite substituents (X's) at
    indefinite positions (Z's)
Ring (R) and Acyclic (A) declarations
Indefinite bonds - dotted
Apostrophe - restricts substitution
Carbon dot - prohibits substitution
Valence declarations
Charge declarations
Boolean Logic (AND, OR (inclusive), BUT NOT)
Repeating units
Abbreviations (Any, Nonc, Hal, Peptides)
Limited retrieval volume 500, substructure search

Figure 15.   CSIS search variables

L-590,226-00A
$C_{19}H_{16}ClNO_4$ Mol. Wt. 357.796
NS
WITZEL, B. M5247-118-5

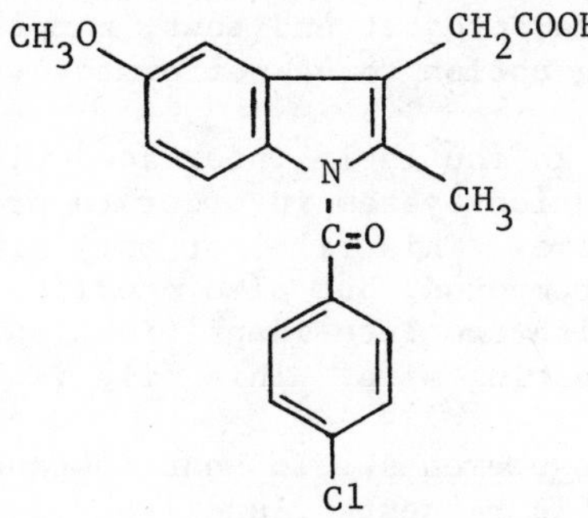

INDOMETHACIN

L-NUMBER	SALT	C L	BATCH	TEST NUMBER	DESCRIPTION
590,226	00	A		2135	ANALGESIC IN RATS 2175
590,226	00	A		2135	ANALGESIC IN RATS 2176

BATCH	LOCATION	CONTAINER
39	TRAY-0242	BOTTLE-036
47	TRAY-0249	BOTTLE-017

Figure 16. CSIS computer report

L-590,226 00A   INDOMETHACIN

L-590,226 04J   N,N*-DIBENZYLETHYLENEDIAMINE SALT OF 1-(4-CHLOROBENZOYL)-2-METHYL-5-METHYL-5-METHOXY-INDOLE-3-ACETIC ACID

L-590,226 07R   POTASSIUM 1-(4-CHLOROBENZOYL)-5-METHOXY-2-METHYL-3-INDOLEACETATE

L-590,226 09V   MONOBASIC ALUMINUM 1-(4-CHLOROBENZOYL)-5-METHOXY-2-METHYL-3-INDOLEACETATE

L-590,226 16T   MIX INDOMETHACIN WITH LACTOSE

Figure 17. Chemical name file

the Transaction File Data Base. Figure 19 shows the format of a typical record.

This system has replaced virtually all of the manual record maintenance in the Chemical Data Department. It is rapidly becoming one of the most important systems. The record contains the L-number, date, project and test, sample size, source and requested testing action to be performed. All of these facets are searchable.

This system is routinely interfaced with the Chemical Structure Information System in order to provide compound profile data for users. This file not only gives an historical record for each compound, but also provides computer generated delivery forms with complete identification, which accompany the samples to the testing site. This file is updated daily.

Chemdata Enhancements. In most instances when a substructure search is requested, ancillary data is also required, most notably biological. The original CSIS design required the printing of every structure retrieved from a search regardless of other search criteria. In order to accommodate all search criteria and yet optimize the utility of search reports, several programs were developed which basically provide a print option. This option is invoked at the completion of CSIS searches. In effect, the option is used to restrict the chemical answers printed from a CSIS search to those with or without biological data and/or with or without sample repository data.

Figure 20 illustrates the commands which have been incorporated as part of the CSIS search question.

Transaction File/Inventory. The next major enhancement, currently under development, is the Chemical Sample Inventory System which is part of the operational Transaction File Data Base. This system provides the researcher with an exact amount of compound available from the Sample Repository. Knowledge of this is essential to those responsible for overseeing and selecting of compounds for both chemical and biological testing. Without this capability, the selection process becomes difficult; sufficient supplies may or may not exist. If adequate supplies of a chosen candidate do not exist, the process must be repeated and the net effect is a costly waste of research time and technical manpower. File building was started by net weighing every new compound submitted to the Sample Repository for cataloguing and storage. Concurrently, samples already in storage were weighed and the net weight data added to the Transaction File.

Figure 21 shows how the record looks with this additional data.

The inventory data are recorded as of a specific date. One of our staff members has written a program which now provides current net weight values. This is accomplished by subtracting

COMPOUND LIST AND LOCATION 01/05/78

SEARCH NO. 010377   W. GALL FOR ANTIINFLAMMATORY STUDIES

L-NUMBER	LOCATION	CONTAINER
590,226-00A	TRAY-0752	BOTTLE-049
	TRAY-0752	BOTTLE-014
	TRAY-0752	BOTTLE-010
	TRAY-0752	BOTTLE-048
	TRAY-0752	BOTTLE-008
	SHLF-0011	BOTTLE-022
	SHLF-0011	BOTTLE-015

Figure 18. Compound repository record

INDOMETHACIN
L-590,226-00A
$C_{19}H_{16}ClNO_4$ Mol. Wt. 357.796
NS
WITZEL, B. M5247-118-5

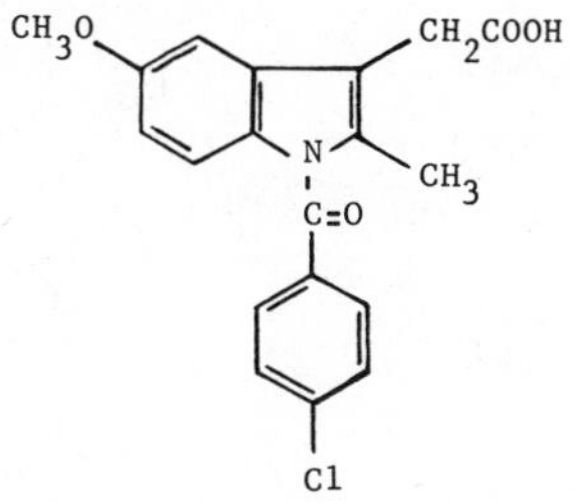

BATCH NO.	PROJECT NO.	TEST NO.	DATE MO DA YR	SAMPLE SIZE	UNITS	SOURCE	COMMENTS
00	0001	0000	01 18 72	0000.1000	GRAM	DC	FAIN BROWN UNIVERSITY, RHODE ISLAND BATCH 104
00	0001	0000	01 18 72	0005.0000	GRAM	DC	SANDBERG, SWEDEN BATCH 104
00	0001	0000	01 18 72	0010.0000	GRAM	DC	ARMSTRONG, CANADA BATCH 104

Figure 19. Transaction file record

B – Print only compounds which do have all the
requested types of biodata.
NB – Print only compounds which have none of the
requested types of biodata.
S – Print only compounds which have sample
collection location.
NS – Print only compounds which do not have
sample locations.

If both biodata and sample codes are present,
a connective "AND" is assumed.

Figure 20.   Print option commands

INDOMETHACIN
L-590,226-00A
$C_{19}H_{16}ClNO_4$ Mol. Wt. 357.796
NS
WITZEL, B. M5247-118-5

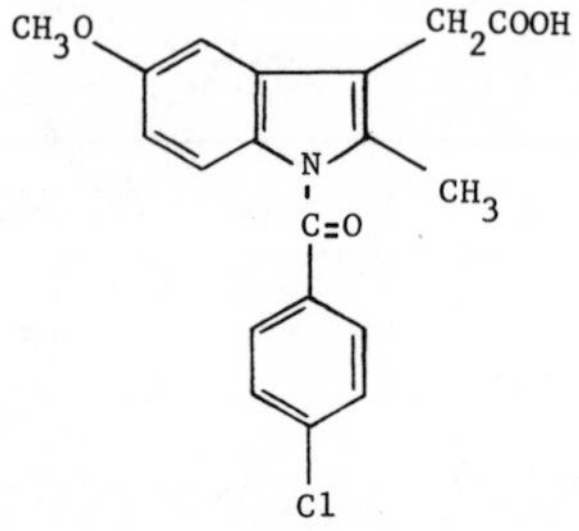

BATCH NO.	PROJECT NO.	TEST NO.	DATE MO DA YR	SAMPLE SIZE	UNITS	SOURCE	COMMENTS
01	0000	0005	10 20 77	0005.1650	GRAM	SC	INVENTORY/DATE 1482-28 ADDITION BY PATCHETT
01	1064	0000	10 26 77	0000.0050	GRAM	SC	DOUGHERTY FOR FE-SOD ASSAY
09	2320	9325	10 26 77	0000.0200	GRAM	SC	DULANEY FOR IN VITRO CHEMOTHERAPY

Figure 21.   Transaction file/inventory report

the weights of samples that have been dispensed from a specific sample after the inventory weight was recorded. This weight is then automatically listed as the "Net Inventory". See Figure 22.

In summary, the "Net Inventory" System coupled with CSIS search retrieval, the transaction profile and biodata results provides the researcher with the desired overview and helps to plan future research activities.

In reviewing the Chemdata Searches completed during 1977 statistics were compiled as shown in Figure 23.

It is obvious for one to conclude that the big use of these systems involves the categories: (1) substructure search with sample availability data, and (2) substructure search with biological and sample availability data.

Interaction

The researcher is using the systems to uncover new pathways of research by analyzing and correlating data that are supplied from the aforementioned files. Given an "idea" about a structure/activity relationship, questions arise:
1. Has compound A been in Test B?
2. If so, what were the results?
3. If not, do we have a sample in the collection? And
4. if so, is there enough material to conduct the test?
These systems are becoming more and more important as they grow and include newer research efforts.

Figure 24 illustrates the interaction among the various files of both biological and chemical data which is now fairly routine at MSDRL.

The first two conditions require examination of the biological data. A positive match with the sample availability file plus a negative match with the structure constraints produce a list of candidate compounds. This list then can be fed back into the Biodata File, the Chemname File, and the Structure File to produce the desired report. Suppose the answer encompasses 150 compounds and 6 of them look promising on this basis. A synthetic program to make analogs may be launched. A new product may result.

INDOMETHACIN
L-590,226 00A
$C_{19}H_{16}ClNO_4$
Mol. Wt. 357.796
NS
Witzel, B. M5247-118-5

01 Net Inventory 5.1400 Gram Tray-1208 Bottle-004

Figure 22.   Net inventory

CSIS Search Totals

1. Total number of searches conducted   695
2. Total number of individual users   86
    A. 74 chemists
    B.   3 patent attorneys
    C.   5 biologists
    D.   4 other information departments

Types of Searches

1. Substructure Types
    A. substructure + sample availability data   216
    B. substructure + biodata   28
    C. substructure + biodata + sample
        availability   313
    D. substructure - compound identification
        only   14
2. Registration Identification
    A. L-number registration + sample
        availability   19
    B. L-number registrations + biodata   28
    C. L-number registrations + biodata +
        sample availability   49
    D. L-number registrations (identification
        only)   28
3. Physical Inventory Searches
    (Operational Nov. 1977)
    A. L-number + biodata + physical inventory   100
    B. L-number + physical inventory

            (2 searches involving 150 compounds)

Figure 23.   MSDRL—use of the Chemdata systems during 1977

Are there any compounds which have:
  An electroshock protection ED_{50} of $\leq$ 50.0 mg/kg
  plus an $LD_{50} \geq$ 200.0 mg/kg, both in mice, and for
  which we have an available sample of at least 150 mg.
Given the above conditions, please display all <u>in vivo</u>
data in computer storage, together with chemical names
and structures, but excluding compounds having 7-membered
rings, with 5 carbons and 2 nitrogens 1,4 to each other.

Figure 24. Biodata/Chemdata interaction to solve query

Acknowledgments

Our thanks go to Mr. T. Boyer, Assistant Director of Merck Sharp & Dohme Research Laboratories Information Systems & Programming, Mr. R. T. Ford, Manager, Research Data Processing, Mr. W. L. Henckler, Senior Chemical Information Specialist of the Chemical Data Department, Mr. W. C. Kinahan, Supervisor, Scientific Information Analyst of the Biological Data Department, Mr. F. W. Landgraf, now of Mobay Chemical Co., Mr. C. J. Miller, Systems Project Supervisor and Mr. W. Pater, Director of Computer Operations.

We also wish to thank many staff members of Merck & Co., Inc. for their part in system development. Particular thanks go to Nancy Nikiper and Ann DeNittis for all the typing involved.

Abstract

An open-ended computer system is described for the collection, storage, retrieval and dissemination of biological data. Facets of the system include variable unit record formats with defined fields derived from significant aspects of test protocols. Interface is commonly made with other data bases mentioned below, singly or in combination, including chemical structure constraints. The common link to the other data bases is the compound registration number. Output display may include records from any or all of the data bases accessed. Counts, results of computation, or tables may be included, as requested by members of the scientific staff.

The Chemical Structure Information System has been partially described in 1976 (2). The complete network of the chemical and biological information systems, including Sample Repository Data Base, Transaction File Data Base, as well as Chemical Name Data Base, is described.

Searches performed during a recent calendar year on Biodata and Chemdata Systems have been analyzed, and the results are discussed with emphasis on the interplay between the chemical structure, substructure and biological data segments of the overall system.

Literature References

1. Emmett L. Buhle, Elinor D. Hartnell, Alexander M. Moore, Louise R. Wiselogle, and F. Y. Wiselogle "A New System for the Classification of Compounds: A Contribution from the Survey of Antimalarial Drugs", J. Chem. Ed., 23, 375 (1946).

2. Horace D. Brown, Marianne Costlow, Frank A. Cutler, Jr., Albert N. Demott, Walter B. Gall, David P. Jacobus, and Charles J. Miller "The Computer-Based Chemical Structure Information System of Merck Sharp & Dohme Research Laboratories", J. Chem. Inf. and Computer Sci., 16, 5 (1976).

RECEIVED August 29, 1978.

Progress toward an On-Line Chemical and Biological Information System at the Upjohn Company

W. J. HOWE and T. R. HAGADONE

The Upjohn Company, Kalamazoo, MI 49001

Over the past ten or fifteen years, researchers in the pharmaceutical industry have seen a gradual but important change in the way computers are utilized in research and research support functions. Early applications tended to focus on numerical tasks such as statistical analyses and quantum mechanical calculations or on the archival storage of information related to the chemistry or biology of research substances. In the latter case, information retrieval systems were often unwieldly and required considerable expertise for their use. The laboratory researcher usually had to work through an intermediary in order to retrieve information from such systems. More recently, we have seen a shift of emphasis to where computers are now recognized as indispensable tools in the day-to-day operation of scientific research. On-line interactive methods have placed the information resource much closer to the end user. In addition to their "traditional" applications, computer-based systems are being employed to assist in the design of organic syntheses, in the interpretation of spectroscopic data, in the design and development of new drug candidates, for real-time experiment control, and in a wide variety of related areas (_1-6_). The retrieval and manipulation of medicinal chemical information is another area in which computer-based systems have made an impact and which will become increasingly important in future years.

This paper will focus on a project which has been under way at the Upjohn Company to develop a comprehensive chemical and biological information system to be used by research scientists and research support personnel. Capabilities of the system will eventually include on-line structure registry, structure and substructure searching, the retrieval and manipulation of pharmacological test data, and the retrieval of spectroscopic, patent, and other types of structure-associated data.

There are currently a number of systems in the company which are being used for the storage of biological data associated with compounds that have been synthesized for screening. In most cases, the operation of these systems has in the past been quite independent of "chemically-oriented" information. Chemical structure

0-8412-0465-9/78/47-084-107$06.25
Published 1978 American Chemical Society

and substructure searching has been accomplished through the use
of a fragment code which was developed in the late 1950's, and
which, despite a number of drawbacks that are commonly inherent
in fragment-based systems, has met the needs of our scientists for
a number of years. However, since one of the goals of the new information system is to provide a means for interactively accessing
both the chemical structure data and associated pharmacological
data, and for the extraction of subgroups of compounds which could
then, for example, act as source data for end-user applications
such as pattern recognition analyses, the design of a flexible and
efficient chemical structure entry and search system became the
initial target of our attention.

The chemical structure system consists of three parts, in
different stages of development:

(a)   the structure database, a collection of approximately 60,000
      chemical structures in connection table format, the construc-
      tion of which has recently been completed,

(b)   the structure entry system, an interactive computer-graphics
      based system which was developed to create the initial data-
      base; portions of this will also be incorporated in the com-
      pound registry and search system,

(c)   the compound registry and search system, currently under de-
      velopment, which consists of two parts:

      (1)   an on-line registry facility which will allow inter-
            active daily updating of the database, and,

      (2)   the query facility, which will allow on-line interactive
            structure and substructure searching and eventual
            searching and manipulation of associated pharmacological
            information. The system will enable the user to display
            the retrieved information in a convenient format and to
            produce high quality hard copy output of both structural
            and textual data.

1. The Structure Database

A key phase of the project involved the creation of the
structure database, a gradually enlarging collection of approximately 60,000 chemical structures which over the years had either
been synthesized in-house for testing purposes or obtained from
outside organizations. The fragment-coded search system also
operated on this collection of compounds; however, since fragment
codes represent structural attributes, the codes could not be used
to regenerate complete connection tables.

A structure entry system was designed which, by using computer graphics as the input medium, would allow direct transcription of the structure diagrams from hard copy format into the computer system. Connection tables would be generated in real-time
as the structure drawing operation progressed. The structure
entry program was ready for use about 1-1/2 years ago and full
scale structure entry began at the start of 1977. Although many

error-detection devices were built into the system, there were still certain types of errors which could slip by and enter the data base. For that reason it was decided at the outset that each structure would have to be entered twice, by different terminal operators, thereby enabling an identity check to be performed on the host computer. Error checking by manual comparison of each entered structure with a hard copy record would, it was felt, take just as long as it would take to redraw a structure a second time and would still provide no guarantee that all errors had been caught.

The structure entry operation has just recently been completed, having taken considerably less time than originally anticipated. Now that the high volume input of the database "backlog" is done, it is planned that routine daily update of the database with low volume "current" structures will be handled by the on-line registry facility which is nearing completion and which will be discussed later.

2. The Structure Entry System

The structure entry system was designed to accommodate rapid error-free structure entry, with much consideration given to structure diagram cosmetics. It was also designed so that it could be easily incorporated into the compound registry and search system with little or no modification. For that reason, we will present an operational overview of the graphical structure entry system, focusing in particular on its use in the creation of the structural database.

(a) Hardware. The data entry terminal is operated essentially as a stand-alone computer system (Figure 1) which transmits completed structure connection tables to the host machine (370/155) where they are compared against their duplicate structures (double entry). Once a day an error log is printed to enable correction of structural errors (using a similar program on the database management terminal, see Figure 1). The structure entry system consists of a PDP 11/04 computer with 28K words of memory, a dual floppy-disk drive, keyboard, graphics tablet, and CRT (similar to the DEC GT43 package). The graphics tablet and associated stylus enable a user to interact with the display by moving the stylus on the surface of the tablet, rather than pointing to the face of the scope as would be done with a light pen. Software in the computer tracks the motion of the stylus with a cursor (a small cross) on the scope. Depressing the stylus activates a switch in the stylus tip which in turn allows the user to select options from a "menu" on the display. Such a device has been found to be a very natural medium for interacting with a display and much more convenient than a light pen. Additional details on the use of a graphics tablet for chemical structure drawing can be found in references *7* and *8*.

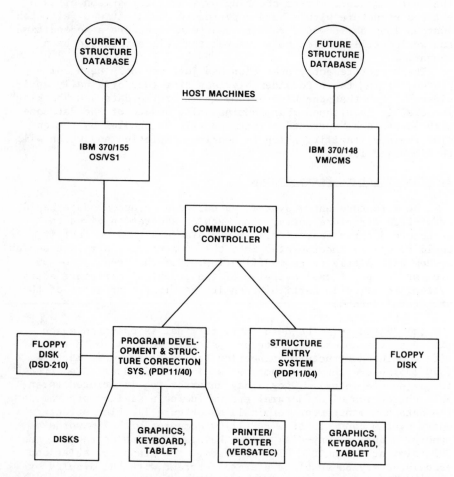

Figure 1. Hardware configuration for structure-entry project. High-volume structure entry was accomplished on the small graphics system (PDP 11/04); data base was formed on 370/155. Data-base maintenance and structure corrections were performed on the large graphics system (PDP 11/40). Information-retrieval-system-runs on 370/148 with data base transferred from 370/155.

At times the host machine is not available, and so rather than transmitting completed structures directly to the database on the 370, the program instead writes them on a floppy disk. These are later incorporated into the database via the floppy disk unit on the second graphics terminal (see Figure 1).

(b) Internal Structure Representation. Structures are represented in the computer in the form of atom-bond connection tables. These are arrays of data which account for such things as:

(i)   for each atom; atom type, formal charge, isotope level, presence of unpaired electron, two-dimensional coordinates, number of bonds attached, and additional information required for regeneration of the structure diagram,

(ii) for each bond; identifiers for the atoms at each end of the bond, bond multiplicity, and stereochemical information for cases where the bond is attached to a chiral atom.

The connection table is formed incrementally during the structure drawing operation. Since X-Y coordinate data for each atom are stored in the table, a complete molecular picture can be generated almost instantaneously from the connection table. The table provides an unambiguous representation of a structure; however, at the time the connection table is inserted in the database, a canonicalization step (using a modified Morgan algorithm (9,10)) is performed which results in a unique ordering of the atoms within the table and facilitates a direct comparison of two "duplicate" tables to detect differences (errors). The connection table that is stored contains no higher-level chemical information such as aromaticity, ring information, or stereochemical relationships other than the bond type transcribed from the hard copy record. Such high-level relationships (and others) can be extracted from the basic information contained in the table by appropriate perception routines on the host machine. In fact, the structure record that will be used for high speed substructure searching is not the original master connection table (CT) for each structure, but a specially formatted record derived from the CT which also contains all the higher-level data necessary to provide search results in as close to "interactive time" as possible (see discussion of substructure searching). Thus, the connectivity information will actually be present in more than one form in the completed registry and search system. In the following discussion, however, "connection table" refers to the expanded connectivity array described at the start of this section.

(c) Graphical Structure Input. Structures are transcribed into the system from data sheets which contain molecular formula, chemical name, structure diagram, some physical and biological screening data, and a registry number called a "U-number". In

some respects, the graphical entry system is similar to those used
by Chemical Abstracts Service (11) for structure input and by computer-assisted synthesis research groups (12,13,1) for specification of target molecules. There are, however, a number of differences from the latter systems due to our required focus on error-
control, speed of entry, and overall structure diagram cosmetics.

A number of drawing options appear on the display, which essentially represent a "menu" of graphics controls. To select from
the menu the user moves the stylus on the tablet so as to superimpose the tracking cursor on one of the options, and then depresses the stylus slightly to activate the desired option. As
can be seen in Figure 2, at the top of the display a rectangle
appears around the TYPE option. This indicates to the operator
the option that is currently active.

Some information must be entered via the keyboard. This includes the date and the operator's initials (at the start of a
session), and a U-number and molecular formula (MF) before each
structure is drawn. The system matches the MF against the structure when the OUTPUT option is selected and only transmits the
structure to the database if the MF and structure match.

The large rectangle in the center of the display represents
the drawing area inside which the molecular diagram is constructed. Error messages and other textual feedback to the user appear
at the bottom of the drawing area.

The options which are arrayed along the top of the display
allow the user to change drawing modes. They operate as follows.
DRAW allows the user to perform a freehand drawing operation to
enter bonds and implicit carbon atoms (see below for description);
RINGS changes the display to a second menu from which pre-drawn
ring systems can be selected; MOVE enables the user to adjust the
position of atoms and their attached bonds by simply superimposing
the cursor on the desired atom and moving the stylus (and thereby,
the atom) to its new position; CENTER centers the drawing in the
box; DELETE allows the selective erasure of atoms or bonds; TYPE
returns control to the keyboard; OUTPUT sends a completed structure to the host machine after the molecule is subjected to a
series of error checks (remaining errors are detected on the 370
during the duplicate match); and CLEAN erases the drawing area and
initializes the connection table.

The three remaining options at the top of the display are for
bond character modification. For example, the broken/zigzag line
allows specification of stereochemical information. While the
system is in this mode, the user can "point" to the center of a
bond and it will become a dashed bond to indicate a projection of
the bond back into the plane of the drawing. Pointing to the bond
a second time converts it to a "wavy" bond of the type normally
used to indicate undefined absolute configuration at a chiral
center. So far, this has been sufficient to permit an adequate
specification of stereochemistry; however, in the next version of
the graphical entry system wedge-shaped bonds will be specifiable,

to increase the flexibility of the system and prevent any ambiguity of chiral site definition.

The arrow option above DELETE is used for the specification of strongly polarized bonds where there is a formal charge separation between the ends of the bond. This can be used in N-oxides or phosphates, for example. The registry and search system will recognize the equivalence of $R_3N \longrightarrow O$ and $R_3N^+ \longrightarrow O^-$, so the arrow is used mainly for cosmetic purposes without any loss of structural information. And finally, the solid line at the top of the display is used to convert any of the special bond types just described back to a normal single bond.

Along the bottom of the display appear a number of commonly-occurring atom types and functional groups, as well as some control options. The FLIP option changes the bottom menu to reveal an additional set of less commonly-occurring atoms and groups. Functional groups not present in preconstructed form can be drawn simply by selecting the component atoms from the menu and connecting them with the appropriate bonds. This, however, takes longer than it does to insert one of the predrawn groups. Many of the predrawn functional groups can also be converted to structurally similar groups. For example, to draw a trichloromethyl group, the operator would (a) insert a CF_3 from the menu, (b) pick up a Cl from the menu and superimpose it on the F_3 in the trifluoromethyl group, and (c) depress the stylus. This would immediately convert the CF_3 to a CCl_3. Since the CCl_3 is represented in the connection table as three distinct chlorine atoms attached to the same carbon, the operator could also enter the same group by drawing the three chlorines separately. Although the appearance would be different, the connectivity data for the two forms would be the same.

At the start of each structure drawing operation, the computer requests the U-number and molecular formula of the compound. After this information is typed in by the operator, the tablet is activated and the picture drawing stage can begin. Although there is no drawing order imposed on the operator, the cyclic nucleus of the molecule is usually drawn first. Rings can be drawn in two ways, freehand or by selecting a predrawn ring system from the second display.

To draw a bond "freehand", the user selects the DRAW option and then depresses the stylus with the cursor inside the drawing area. As the stylus is moved, a straight line appears on the scope as if it were "ink" from the stylus. When the stylus is lifted, the line is frozen and the new bond is inserted in the connection table. The terminating atoms are initially assumed to be carbon. Additional bonds can be drawn in this manner until the desired ring is formed.

We have found, however, that the entry operation can be speeded up considerably by providing a collection of predrawn ring systems which can be brought into view (Figure 3) by pressing the RINGS option.

Figure 2. Structure-entry display showing some of the graphics options available

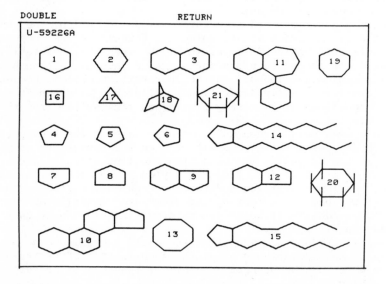

Figure 3. Structure-entry display showing predrawn ring systems available

This feature provides an additional benefit in that most of the resulting rings are regular, have similar edge lengths, and can be linked together quite easily. To pick up one of the rings, the operator depresses the stylus on the numeric symbol inside the appropriate ring. The display changes back automatically (to Figure 2) and the selected ring picture becomes the new stylus cursor, following the stylus motion.

The user moves the ring to the appropriate position on the screen and depresses the stylus again. This action causes the ring atoms and bonds to be inserted in the connection table and the ring to be frozen at that position on the scope. But since the original ring is still tracking the stylus, additional rings can be inserted, either separately or linked together. The program automatically takes care of shared bonds (fusion) or shared atoms (spiro or bridging systems) as ring networks are built up during the ring linking operation (Figure 4). The ring is released from the stylus tip and the original cursor reappears as soon as the ring cursor is moved outside the drawing area.

We have found that this allows the operator to build complex ring systems quite rapidly and eliminates the many diagram adjustments (using the MOVE option) that would otherwise have to be made.

To insert non-carbon atomic symbols, the user depresses the stylus on the desired atom at the bottom of the display. The tracking cursor disappears and is replaced by a copy of the selected atomic symbol. This can then be moved into position in the developing molecule by depressing the pen at the appropriate position, causing a new atom to be created or a previously existing atom to be converted to the new atom type. Attached bonds are automatically shifted back slightly to make room for the new atom symbol.

Groups of atoms and functional groups can be inserted in a similar manner. When a functional group is selected from the menu and positioned in the developing molecule, the program does not simply adjust the picture to display the new group. Rather, the system modifies the connection table to include the new information. In the case of a nitrile group, for example, carbon, nitrogen, and triple bond entries are automatically inserted in the table. The picture is then regenerated from the updated CT. In fact, any modification of the structure is accompanied in real-time by a corresponding connection table update.

As each drawing operation is performed, the program executes a valence check. Thus, the attachment to an atom of excess bonds or the conversion of a fully-bonded atom to an atom type of lower valence is prevented and accompanied by an appropriate message to the operator. Insertion of a charge on an atom (via the "+" or "-" options at the bottom of the display) allows attachments in excess of the normal valence to be made.

Acyclic atomic chains can be drawn by picking up the constituent groups and atoms from the menu, inserting them in the

drawing, and then bonding them together using the DRAW option.
However, this approach tends to be slow and connecting bonds can
be left out inadvertently (this condition would be identified by
the error check when the OUTPUT option is pressed).

The user instead generally selects one of the four arrow options from the bottom of the display (Figure 2). When the stylus
is then depressed on an atom, a blinking arrow appears beside the
atom, pointing in the direction of desired chain extension. As
long as the blinking arrow remains on an atom, groups or atoms can
be automatically inserted and bonded with even spacing simply by
depressing the stylus on the desired atom or group option at the
bottom. By contrast, with the arrow off, selection of a group or
atom results in the cursor being replaced by a copy of the group
or atom symbol; the user must then move it up to the correct position for insertion. By this "automatic insertion" method, acyclic
chains can be constructed quite rapidly with a minimum of hand
motion.

Sequence *I-VII* depicts the various stages of chain growth for
a sample molecule. As each atom or group is added, the blinking
arrow moves over (assuming no valence infraction resulted from the
insertion) to remain at the end of the developing chain. If the
group with which the arrow is associated is a $(CH_2)_2$ unit, depressing the (CH_2) option causes the subscript to be incremented
(up to a maximum of $(CH_2)_{30}$), while the arrow stays in place. The
connection table is updated to reflect the fact that there are now
three connected methylene units (in structure *VI*) with the same
coordinates. When a group is added which has no remaining free
valence (such as a nitro or methoxy), the arrow turns off. At any
time the arrow can be turned off or pointed in another direction
by selecting any of the other options on the display, including a
different arrow option. Figure 5 illustrates a molecule being developed with the assistance of the blinking arrow option.

The program also provides the capability for handling compounds with more exotic structural relationships, such as those
exemplified in *VIII-X*.

Compound *VIII* represents a mixture of two moeities in a 200:1
molar ratio. Structure *IX* has a non-integral number of moles of
methanol and an undetermined number of moles of water solvated to
the main structural unit. Compound *X* represents a pure compound
with unidentified stereochemistry at the asymmetric double bond.
Structures which contain incompletely defined positioning of ring
substituents, for example, could also be handled in a manner similar to *X*. In a newer version of the graphical entry system being
designed for use in the registry and search system, provision is
being made for polypeptides, polynucleotides, metallocenes, and
metal coordination complexes which cannot currently be handled.

When the entry of a structure has been completed the operator
depresses the stylus on the OUTPUT option. The software checks
for missing bonds (unfilled valences on atoms for which attached
hydrogens are not implied), unbalanced atomic charges, and mis-

Figure 4.   *Illustration of ring-network construction*

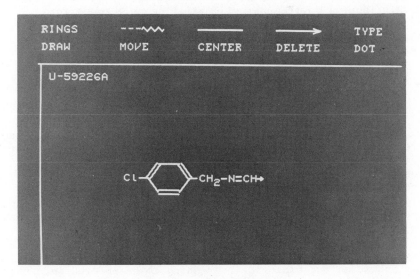

Figure 5.   *Automatic chain extension with arrow option (arrow is on CH)*

matches between the MF and the structure (Figure 6). If an error
is detected, a message is displayed at the bottom of the screen
explaining the problem. If possible, the program will identify
the molecular site at which the error exists. In the example illustrated in Figure 6 the molecular formula was incorrect and
would have to be retyped. A number of errors that have existed
even on the original data sheets have been identified by this approach and corrected. The finished structure is transmitted to
the database on the 370 where any remaining errors such as mis-
placed substituents or inaccurately specified stereochemistry are
flagged by the duplicate checking operation for later correction.
Only in the unlikely event that two operators make identical errors of the two kinds just mentioned will incorrect structures
actually enter the permanent database.

     As is evident from the above discussion, considerable impor-
tance has been attached to the visual appearance of the structure
diagrams as they are inserted in the database. Among the many
reasons for this are: (a) the planned inclusion in the registry
and search system of a report generation facility (high quality
structure plots are already being generated from the database for
insertion in the data sheets of "current" compounds), and (b) the
desire for neat, unambiguous, and consistent (within groups of
structural analogs) displays of compounds retrieved by users of
the registry and search system. It should be stressed, however,
that when a user draws a structure or substructure into the system as the target of a search, there need be no concern whatever
for structure diagram cosmetics. Structures can be drawn using
the freehand techniques described earlier, with convoluted and
uneven rings, for example, without affecting the accuracy of the
search. Thus, a novice user will still be able to perform
searches without having to learn the fine points of ring network
construction, automatic chain lengthening, functional group
interconversion, and so on.

3. The Compound Registry and Search System

     The compound registry and search system provides the graph-
ical user interface, interactive command language, search functions, and database management support which allow compound-asso-
ciated information to be added to the database and existing data
to be modified or queried. The registry facility is available on
a restricted basis to those users with authority to update the
compound database. On the other hand, the query facility is
available to all users and allows search, display, and hard copy
output of the various types of information stored in the database.

     (a) The Registry Facility. The principal function of reg-
istry is to allow chemical structure information on compounds
newly submitted for biological screening to be entered into the
compound database. Although full operation of the registry por-

tion of the system is still several months away, the design has
been completed and implementation is now in progress. The major
components of the facility include:

(a)   a graphical structure entry and display subsystem em-
      bodying all the features of the structure entry package
      just discussed,

(b)   a command language, interpreted by the host processor,
      and

(c)   a variety of search, update, and deletion functions.

Commands entered at the keyboard of the graphics terminal
appear on the scope and are simultaneously transmitted to the 370
for interpretation. If an authorized user wishes to register a
new compound, the host machine signals the graphics system to activate the drawing controls discussed earlier. The user may
either draw the complete molecular diagram or modify a previously
fetched structure which was known to be similar in appearance to
the new compound. The latter capability results in faster structure entry since much of the drawing operation is avoided. When
the drawing is complete, a number of actions are performed by the
host machine prior to actual insertion of the new compound in the
database. First, an exact structure search (taking on the order
of 2 seconds) is performed over the entire database. The internal execution format of the structure search will be described
briefly in the section on implementation. At the user's option,
the search may be performed with or without regard to stereochemical differences. A "hit" resulting from the structure
search indicates that the target compound is, in fact, not new
and no registry takes place.

"Fragmented" compounds such as salts, solvates, mixtures of
two or more components, and metal coordination complexes pose a
slightly different problem which is based upon our traditional
assignment of the same U-number to all salts, for example, in a
particular structural series, with letter suffixes appended to
the U-number to differentiate the individual salts (or solvents,
or coordinated metals, etc). A secondary search capability is
provided which allows the user to select the structural subunit
upon which the fragment search is to be based. The system then
retrieves the U-numbers of all compounds which contain the selected fragment. The user can assign a registry number to the current compound based upon his examination of the retrieved structures. To illustrate, a registry attempt on structure *XI* in
which the carboxylate-containing unit was specified as the fragment of interest would retrieve compound *XII* (the retrieved
structures are displayed beside the original structure to facilitate visual comparison). The system also converts the search
fragment to an unionized acid form and would retrieve *XIII*.
Registry of *XI* then involves the assignment of the same base number (as *XII* and *XIII*) plus a unique suffix.

The registry of new compounds is accompanied by the same
error checks that were discussed for the data entry system, but

I — select arrow, place on atom → II — select CH_2 → III $-CH_2 \rightarrow$

select $\overset{H}{N}$ → IV $-CH_2-\overset{H}{N} \rightarrow$ select (CH_2) → V $-CH_2-\overset{H}{N}-(CH_2)_2 \rightarrow$

select (CH_2) again → VI $-CH_2-\overset{H}{N}-(CH_2)_3 \rightarrow$ select NO_2, arrow turns off → VII $-CH_2-\overset{H}{N}-(CH_2)_3-NO_2$

VIII

200   and   1

IX   • 2.5 CH_3OH • x H_2O

X   or

XI

XII

XIII

since the double entry approach is no longer used, the visual inspection of structures plays an important role in compound registry. Additional features of the registry facility provide the
capability for modification and deletion of existing data.

(b) The Query Facility. The query facility provides the
general user a means of obtaining information from the compound
database. A major goal in the design of the user language has
been to make it simple enough for the inexperienced user to approach and yet powerful enough to allow the more knowledgeable
user to formulate complex queries. To meet this need we have designed two languages. The first language, called the novice language, is aimed at the inexperienced or occasional user and is
simple enough to allow him to retrieve information without the
assistance of an information specialist. At each point during
the interactive dialog the user is presented a menu of available
options. At any point he may enter the "HELP" command to receive
a more extensive description of the various options. For purposes of simplicity a limited set of queries is available in novice mode.

The second language, called the expert language, has been
designed for the more experienced user. It differs from the novice language in that the user is assumed to have a working knowledge of the syntax and semantics of the language. Queries can be
stated in a compact manner and a fuller range of requests may be
specified. Additionally, it has been designed to allow future
data element types and search capabilities to be integrated into
the language in a natural manner. The following examples illustrate some of the types of queries that can be stated with the
expert language.

SEARCH COMPOUNDS WHERE STRUCTURE = S1

In this example the SEARCH command is being used to search
for an exact structure. In response to this command the structure drawing graphics controls are displayed and the user is
asked to draw the structure. Structure specification is performed by means of a graphics subsystem similar to that of the
structure entry system. The user has the additional option of
specifying absolute configurations or leaving stereochemistry undefined. In the latter case more than a single structure match
could occur. Once structure input has been completed, the search
is executed and the user is notified of the number of compounds
that matched the target structure. The set of registry numbers
of the matching compounds is written into the "search results
file", a temporary holding area that can be manipulated by other
commands to print and display associated information, or which
may be saved for later reference.

SEARCH COMPOUNDS WHERE STRUCTURE > F1 OR STRUCTURE > F2

　　　This query specifies a SEARCH for compounds that contain at
least one of two exact structural fragments. The term "fragment"
is used here to refer to a complete connected structural unit.
Such a search will find all salts, solvates, and mixtures in the
database that contain a particular component much more quickly
than will a general substructure search. In response to the command the structure drawing controls are displayed and the user is
asked to draw the structural fragments F1 and F2, in turn, The
search is then performed and the registry numbers of qualifying
compounds are written into the search results file. In this example a disjunction of the two predicates is specified; in addition, general Boolean expressions involving conjunctive and nega-
tion operators may be formed and parenthesis pairs may be used to
control operation hierarchy.

SEARCH COMPOUNDS WHERE ENTERER IN (RCA, EN, CD) AND ENTRYDATE
BETWEEN 1-DEC-77 AND 1-JAN-78 AND U# > U-55,000 AND STRUCTURE
> SS1

　　　To qualify for retrieval in this SEARCH example a compound
must pass four tests: First, the person that entered the compound originally must have the initials RCA, EN, or CD. This is
an example of the set predicate which allows a set of qualifying
values to be listed within a pair of parentheses. Second, the
compound must have been entered between the two indicated dates.
This is an example of the range predicate in which lower and upper inclusive qualifying bounds are specified. Next, the registry number must be greater than U-55,000; and finally, the compound must contain a substructure. The substructure drawing controls are displayed and the user draws the search target. Substructure specification has elements in common with structure
specification and, in addition, provides some special constructs
for defining substructures. The user may indicate variable substituents, indefinite positioning of groups, and variable-sized
rings and chains *(14)*. The more complicated substructure definitions will require the input of both graphical and textual data.

DISPLAY STRUCTURE, MF WHERE ENTRYDATE = 1-JAN-78

　　　The DISPLAY command is used to fetch and display selected
information at the terminal. In this case the registry number,
structure diagram, and molecular formula of all compounds registered on the first day of 1978 are displayed. Any subset of the
stored data fields may be displayed for each compound and the
screen is formatted in the most convenient fashion for the fields
selected. To see the next screenfull of information the user depresses the "return" key; to terminate the display sequence the

STOP command is used. The WHERE clause of the DISPLAY command is
of the same form as in the SEARCH command and may include non-
structural data as well as structure and substructure search requests. The SEARCH and DISPLAY commands differ in that, while
both search the database, save the qualifying registry numbers,
and present a "hit count" to the user, the DISPLAY command also
displays selected information at the terminal for each compound.
In addition, DISPLAY can operate on the set of registry numbers
retrieved by a previous search.

PLOT STRUCTURE FORMAT 4 WHERE U# > U-55,000

The PLOT command is used to produce hard copy output of
selected information in a manner analogous to the DISPLAY command.
In this example a request is made for registry numbers and structure diagrams of compounds with registry numbers greater than
U-55,000, plotted four per page. Plots are produced on a Versatec
electrostatic printer/plotter (all structure diagrams in this paper were produced with the structure plot facility). Both structural and nonstructural information can be produced together in a
variety of formats.

SAVE FILE1

This command saves the registry numbers retrieved in the most
recent search in a file named FILE1. Saved search results may be
accessed at a later time by any of the commands described above
and may be combined with other saved search result files in new
queries. Each user may have many saved search result files which
can be kept indefinitely.

(c) Implementation Details. All structural and nonstructural data will be stored and retrieved through the facilities of
a relational database management system (RDBMS) *(15,16,17,18)* with
the exception of specially-formatted connectivity information used
in conjunction with the substructure search subsystem. The relational approach provides a simple data model in which information
is organized in the form of tables. A high level query language,
similar to the expert mode query language, frees the programmer
from dealing with the low level physical storage structure of the
data. This "data independence" allows data structures to be
changed and indices to be created and destroyed to meet the
changing performance requirements of the system without necessi-
tating program modifications. In addition, an RDBMS will significantly reduce the system implementation time, compared with that
required by traditional techniques.
The search request, as received from the user, consists of
three basic types of search elements: nonstructural data, exact
structure data, and substructure data. The syntax parser classi-
fies each element of the search request into one of the three

categories. A search is then performed over the database for
each category and the retrieved registry numbers are combined to
produce the final search result.

To execute the portion of the search involving only nonstructural data, such as biological or patent information, the system
performs a relatively straightforward translation of the user's
request into the query language of the RDBMS, which supports all
of the predicate and logic constructs available with the expert
language. The RDBMS then executes the search and returns the set
of registry numbers of qualifying compounds.

The portion of the search that involves exact structures and
exact structural fragments is executed in two phases. First an
8-byte hash value is calculated from the canonicalized *(10)* connection table for each exact fragment and structure. The query
facility of the RDBMS is then used to search the database for all
compounds that contain structures or fragments with the calculated hash values. Next, the connection tables for qualifying
compounds are retrieved and the relevant attributes for the atom
and bond descriptors are compared to determine if the structures
do, in fact, match the target structure.

The portion of the search which involves chemical substructures presents more difficulties. Since a general-purpose database management system cannot be employed efficiently to perform
substructure searching (SS), a special approach is required. The
chemical information literature is replete with examples of the
effort that has been directed toward the development of efficient
SS algorithms. Good screening techniques have been developed,
(19,20,21) set reduction methods have evolved, *(22,23)* and atom-
by-atom search techniques have been refined *(14,24)*. More recently, increasing attention has been given to the use of specialized hardware in SS *(24)* and in the closely related problem
of textual database searching *(25)*.

The traditional approach to SS has been the "three phase"
search. The first step involves some form of full structure
screen in which from several hundred to a thousand or more pre-
calculated screen bits are used to quickly eliminate 90 to 99.9
percent of the structures. This is generally followed by a step
in which candidate atoms are located in the remaining structures
for each substructure atom; and finally, an atom by atom mapping
of the substructure into the structure is performed.

We have decided to take an approach in which specialized
hardware is employed to perform a sequential scan of the database
in conjunction with a moderately efficient screen. The screening
is performed in real time by an intelligent disk controller as the
data passes the read head of the disk. This is followed by a
very efficient atom and bond candidate selection step and then a
final atom-by-atom match. These last two steps are performed on
a dedicated minicomputer. The general component configuration of
the SS system is shown in Figure 7.

When the search request is received by the host machine the

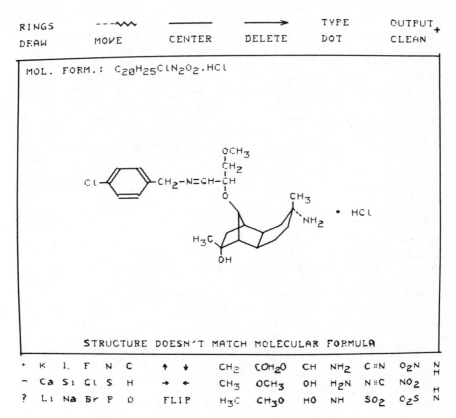

Figure 6.   *Illustration of error detection prior to insertion of structure in data base. Message at bottom of drawing area says "structure doesn't match molecular formula."*

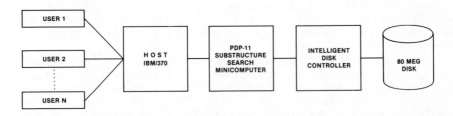

Figure 7.   *General hardware-component configuration of substructure-search system. Front end consists of graphics minicomputers. Back end consists of dedicated minicomputer, "intelligent" disk controller, and dedicated disk.*

SS portions are extracted and sent to the SS minicomputer for execution. The SS minicomputer then sends the substructure screen
bits to the intelligent disk controller and instructs it to start
scanning the disk. The structure screens and connection tables
will be stored on the disk in the format shown in Figure 8. A
ones-complement operation is performed on the structure screen
before it is written on the disk. Therefore, bits with a value
of 1 represent those structural attributes that are absent in the
structure. As the screen bits of each structure pass the read
head of the disk they are read by the controller and logically
AND'ed with che substructure screen bits supplied to the controller by the minicomputer. If the result of this operation is non-
zero the structure cannot possibly contain the substructure and
is eliminated from further consideration; otherwise, the connection table is read into the main memory of the minicomputer for
further processing. The disk is read in a sequential manner and
when the end of a track has been reached the disk head is stepped
over to the next track. Scanning continues after a one revolu-
tion delay. While the disk is being scanned by the controller
the minicomputer is simultaneously executing the candidate selection and atom-by-atom matching portions of the search.

The atom and bond candidate selection step is performed by
an algorithm that combines bit screen and set reduction techniques. The connection table is arranged in a special format,
with one table entry for each bond in the structure. Each entry
contains the atom types and sequence numbers of the atoms at each
end of the bond as well as the bond type. Entries are ordered by
increasing frequency of occurrence (based on statistics calculated over the entire database) of the simple pair (atom-bond-atom
sequence) containing the bond. In addition, a small number of
screen bits, called a pair screen, is associated with each bond.
The pair screen, which is a function of atom and bond sequences
within a radius of 2 bond lengths of the central bond, describes
the structural environment in the immediate neighborhood of the
bond in a manner similar to that of a full structure screen. The
pair screen bits are calculated at the time the compound is registered and are stored permanently in the database. Although
these extra bits increase the size of the database, experiments
have shown that they help provide short and relatively consistent
search times.

Execution proceeds by selecting, in turn, each entry in the
substructure table and screening against it those entries in the
structure table that are of the same simple pair type. The com-
plemented screen bits of each qualifying structure entry are logically AND'ed with the screen bits of the substructure entry in
the same manner as for the full structure screen described above.
A result of zero indicates that the environment of the selected
bond in the structure is similar to the environment of the current bond in the substructure. Candidate information is stored
for each structure bond that matches the substructure bond, to be

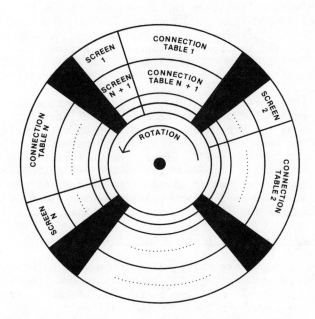

Figure 8.   Layout of structures on the disk for substruc-
ture searching

used later in the final atom-by-atom mapping of the substructure
into the structure. If any substructure atom or bond fails to
have a candidate in the structure, the examination of that structure is halted (a "no match" condition).

Figure 9 shows an example of the candidate selection process
utilizing a simplified pair screen of four bits per bond (although
the optimal number of screen bits has yet to be determined, it
will be in the range of eight to sixteen bits per bond), which
represent an adjacent single bond, an adjacent double bond, an
attached oxygen atom, and an attached carbon atom. In this case,
the following occurs: the third structure entry is screened
against the first substructure entry (same simple pair type) and
passes the screen; the last two structure entries are screened
against the second substructure entry and only the fourth structure
entry passes the screen; and finally, all of the structure entries, except the third, are screened against the third substructure entry and only the first passes. The arrows in this figure
indicate the structure bonds to which each substructure bond has
been mapped.

If a structure passes the candidate selection step, an atom-
by-atom mapping of the substructure into the structure is performed and the registry numbers of compounds that qualify are returned to the host machine as they are found. Since the SS system
has yet to be implemented in final form, accurate SS performance
data are not available; however, time projections, based on current disk technology and an already implemented SS prototype system, indicate that most searches will require about 30 seconds
elapsed time for the 60,000 compound database.

4.   Integration of Biological Data: Future Goals

Although much work still needs to be done before the compound
registry and search system will be operated on a routine basis,
most of the difficult problems concerning chemical structure
handling have been overcome. In the next major phase of the project the principal effort will focus on "biological data", a term
which encompasses a very broad range of information in the field
of pharmacological studies. The biological data handling capabilities of the query system will undergo a continuing evolution
which will come about not only as new types of pharmacological
data become available for incorporation into the system, but also
as the need for (and availability of) new techniques for manipulating experimental data evolves.

Initial work on the incorporation of biological information
into the compound registry and search system will deal mainly with
data that is already being captured on a routine basis for computer input and storage. This includes screening results in which
the biological response of compounds to a variety of test screens
is indicated by numerical activity values or binary activity as-
signments (active/inactive). Additional data types to be incor-

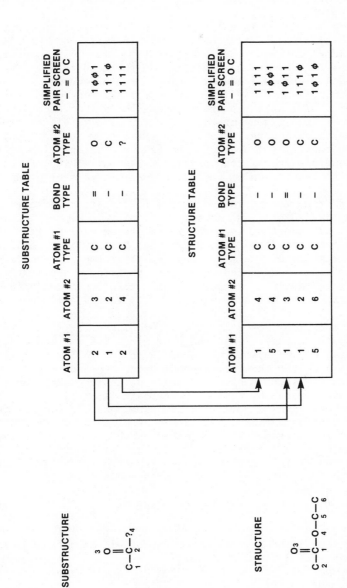

Figure 9. Simplified candidate-selection example (this is the second phase in a substructure search)

porated will eventually include toxicity information and more detailed activity results that pertain to individual classes of
pharmacological agents.

As was mentioned in the section on implementation, searches
over the biological portion of the database will be controlled by
the relational database management system. The logic constructs
of the expert query language will allow the user to specify
rather complex chemical and biological search requests in which,
for example, the database is searched for all compounds that contain a particular substructure, which also exhibit a desired activity level in a given screen, and which also were submitted
after a particular date. Use of the RDBMS promises not only to
reduce substantially the effort required for integration of the
chemical and biological databases, but also will simplify considerably the evolution of biologically-oriented search capabilities.

In addition to providing a means for interactive searching
of chemical and biological data (for display or report generation
purposes), an important feature of the system will be its ability
to create subsets of the main database. Users will be able to
treat the results of their searches as their own private databases which can be accessed by specially tailored application
programs. For example, compounds which were retrieved by a combined substructure and screening activity search could become
source data for more detailed analyses using pattern recognition,
molecular modelling, or statistical techniques.

Although we expect that the biological information handling
capabilities of the system will undergo a continuing evolution,
there is a need for the inclusion of other types of data as well.
Spectral data, patent status information, CAS registry numbers,
chemical names, and physical property data all fall under the um-
brella of "medicinal chemical information" and are some of the
more important data types that have been planned for eventual inclusion in the system. The projected capabilities of the system,
enabling a user to interactively query and manipulate such
diverse types of information, should make the system an important
asset in the research and research management functions.

Literature Cited

1. *Computer-Assisted Organic Synthesis*, Wipke, W. T. and Howe,
    W. J., eds., ACS Symposium Series No. 61, American Chemical
    Society, Washington, D.C. (1977).
2. *Minicomputers and Large Scale Computations*, Lykos, P., ed.,
    ACS Symposium Series No. 57 (1977).
3. *Computer-Assisted Structure Elucidation*, Smith, D. H., ed.,
    ACS Symposium Series No. 54 (1977).
4. *Chemometrics: Theory and Application*, Kowalski, B. R., ed.,
    ACS Symposium Series No. 52 (1977).

5. *Algorithms for Chemical Computations*, Christofferson, R. E., ed., ACS Symposium Series No. 46 (1977).
6. *Computer Networking and Chemistry*, Lykos, P., ed., ACS Symposium Series No. 19 (1975).
7. Howe, W. J. and Hagadone, T. R., "Substructure Searching", in *Proceedings of the Technical Information Retrieval Committee of the Manufacturing Chemists Association*, Washington Meeting, 1977, in press.
8. Corey, E. J. and Wipke, W. T., *Science*, <u>166</u>, 178 (1969).
9. Morgan, H. L., *J. Chem. Doc.*, <u>5</u>, 107 (1965).
10. Wipke, W. T. and Dyott, T. M., *J. Amer. Chem. Soc.*, <u>96</u>, 4834 (1974).
11. Blake, J. E., Farmer, N. A., and Haines, R. C., *J. Chem. Inf. and Computer Sci.*, <u>17</u>, 223 (1977).
12. Corey, E. J., Wipke, W. T., Cramer, R. D., and Howe, W. J., *J. Amer. Chem. Soc.*, <u>94</u>, 421 (1972).
13. Wipke, W. T., in *Computer Representation and Manipulation of Chemical Information*, Wipke, W. T., Heller, S. R., Feldman, R. J., Hyde, E., eds., p. 147, Wiley Publ., New York (1974).
14. Brown, H. D., Castlow, M., Cutler, E. A., Jr., Demott, A. N., Gall, W. B., Jacobus, D. P., and Miller, C. J., *J. Chem. Inf. and Computer Sci.*, <u>16</u>, 5 (1976).
15. Codd, E. F., "A Relational Model of Data for Large Shared Data Banks", *Commun. of the ACM*, XIII, 377 (1970).
16. Codd, E. F., "Further Normalization of the Data Base Relational Model", *Courant Computer Science Symposia 6, Data Base Systems*, Prentice-Hall, New York (1971).
17. Date, C. J., *An Introduction to Database Systems*, Addison Wesley, Reading, Mass., (1975).
18. Astrahan, M. M., *et al*, "System R. Relational Approach to Database Management", *A.C.M. Transactions on Database Systems*, <u>1</u>, 97 (1976).
19. Feldman, A., Hodes, L., *J. Chem. Doc.*, <u>15</u>, 147 (1975).
20. Adamson, G. W., Bush, J. A., Mclure, A., and Lynch, M. F., *J. Chem. Doc.*, <u>14</u>, 44 (1974).
21. Meyer, E., "Superimposed Screens for the GREMAS System", in *Proc. FID-IFIP Conference*, p. 280, Samuelson, K., ed., Rome Meeting, 1967, North Holland Publ. (1968).
22. Sussenguth, E. H., Jr., *J. Chem. Doc.*, <u>5</u>, 36 (1965).
23. Figueras, J., *J. Chem. Doc.*, <u>12</u>, 237 (1972).
24. Haines, R. C., "Substructure Search Design Study Status Report", Chemical Abstracts Service Working Paper (unpublished), 1976.
25. Bird, R. M., Tu, J. C., Worthy, R. M., "Associative/Parallel Processors for Searching Very Large Textual Data Bases", SIGIR-SIGARCH-SIGMOD Third Workshop on Computer Architecture for Non-numeric Processing, McGill, M. J., ed., *SIGMOD*, <u>9</u>, No. 2, 8 (1977).

RECEIVED August 29, 1978.

9

Warner-Lambert/Parke-Davis-CAS Registry III Integrated Information System

ROGER D. WESTLAND, RAYMOND L. HOLCOMB, JOHN W. VINSON,
JON D. STEELE, ROBERT J. CARDWELL, ROBERT L. SCOTT,
THOMAS D. HARKAWAY, PATRICIA J. HYTTINEN, and TINA WILLIAMS

Warner–Lambert/Parke–Davis Pharmaceutical Research Division,
Ann Arbor, MI 48105

In 1946 the Parke-Davis Research Laboratories centralized chemical and biological research data using manual methods of storage and retrieval. These were effective until the late 1950's, when manual methods were gradually reinforced with punched card files. By the mid 1960's, machine readable data files were available for everything except a complete chemical structure and certain other structure-related information. Throughout the development of computerized information systems it has been necessary to maintain redundant manual files until nearly all information is computer-readable. Only now, after adding chemical structures to the computer database can we begin to abandon the manual files maintained for over 30 years. In addition to structure-handling capability, we have developed a system to link sample inventory and properties, biological screening data, and research document data to produce reports and answers to queries, both interactively and in batch mode.

In considering approaches to computerized chemical structure processing (1, 2, 3), we accepted an offer by Chemical Abstracts Service (CAS) to establish under contract a private satellite of the CAS Registry System (4) which employs over 640 programming modules and over a quarter-million source statements. Since Warner-Lambert/Parke-Davis (WL/PD) had compatible hardware for both processing and structure printing, we were in a position to take advantage of CAS's large investment in high quality graphics, name processing, and computer edits. CAS offered an advanced and highly developed system which could be installed in a short time at relatively low cost. Ongoing development at CAS to enhance the system for storing, retrieving, and reporting the world's chemical literature made compatibility with CAS attractive. Current use of CAS's service in Europe (5), Japan (6), and the United States (7, 8, 9) evidences increasing reliance on the CAS Registry System and suggests the possibility of broad industrial and governmental use in the future.

A pilot project at WL/PD required less than two-man months of effort to implement CAS's structure-printing algorithms from

0-8412-0465-9/78/47-084-132$05.00
Published 1978 American Chemical Society

the CAS Graphical Data Structure (10, 11, 12) record. Success of the experiment in plotting structures of the type shown in Figure 1 stimulated further exploration which ultimately led to the development of a WL/PD - CAS integrated system for storing, retrieving, manipulating, and reporting chemical and biological research data.

System Design

With an INQUIRE® (Infodata Systems Inc., Falls Church, Virginia) database management system available on our IBM 370/168 computer, historical computer files of sample inventory and transactions, physical and chemical properties, biological screening data, research document data, and other miscellaneous files were converted to INQUIRE file format (13), and stored on disk (Figure 2). Sample transactions (to and from physical storage) and inventory data are entered by interfacing with the central computer an on-line balance and a keyboard-CRT terminal. Other WL/PD information is entered in a key-to-disk operation using the ENTREX® (14) system, thereby providing options for direct entry of data from laboratories, when appropriate. Output from the Private Registry files at CAS is converted by means of update programs to INQUIRE file formats. Multi-file searching of the INQUIRE files for ad hoc queries or report construction can be done either interactively with TSO terminals or in batch mode using a Varian V74 computer as a HASP work-station. Generic structure searches of the computer file of fragment-coded structures give as optional output punched paper tape that controls the display of structure images on microfiche. The coded microfiche containing 196 structure images at 24X reduction are stored in the carousel of a storage and retrieval unit manufactured by Image Systems, Inc. Since a new substructure search system for the WL/PD file will not be usable until the entire backlog of structures has been entered into the Private Registry, we are considering programs to algorithmically generate the Parke-Davis Fragmentation Code (15) from CAS connection tables. This will allow us to continue using our present search techniques in the interim.

Properties File. The following data are included in the key-to-disk entry of properties: accession number, source, percent of parent component, melting or boiling point, special handling or storage requirements, physical state, solubility, stability, selected analytical and spectral data, sample weight and location, submission date, and literature references.

Transactions File. A Mettler Model PT320 balance having BCD output, and a CRT terminal are interfaced with the central computer through a microprocessor and the Varian HASP work-station. At the time sample weights are automatically recorded, the operator keys 1) transaction type, 2) accession number, 3) date, 4) whether the sample is being received and from whom, or being transmitted and to whom, and 5) storage location. While this information is stored in the "Transactions" database a running record of the amount of sample on hand is calculated from on-line balance entries and stored in the "Properties" database.

Biology File. Screening data from biology laboratories are recorded on data entry forms appropriately coded (13) for key-to-disk handling, either in a central location or the laboratory itself. Result forms are customized for each test and are rearranged into a standard format by the ENTREX processor before being sent to the main computer.

Document File. Search parameters of internally generated research reports are included in the Document File (16). Text (word) processing equipment, soon to be acquired, will permit inexpensive recording of selected text such as abstracts. A variety of options to INQUIRE include techniques which can index and retrieve on the basis of such text. The multi-file option allows selected records to be combined with data from other INQUIRE files.

CAS Files. Machine processing of data must be performed at CAS to take advantage of the many machine validating and duplicate checking features of the CAS Registry System. Although structures and chemical names could be entered at the user's location followed by transmittal of computer-readable data to CAS for processing, CAS's keyboarding conventions and high volume allow them to offer the service at a cheaper rate than we could match internally. Accordingly, data sheets of chemical structures and names are shipped to CAS on a twice-weekly basis (Figure 3). At CAS the hand-written information is checked and edited, and structures, stereo-descriptors, and names are entered by a key-to-disk procedure (17). Keyboarded records of structures are processed in the Private Registry satellite system with the use of most of the computer edits of the CAS Registry System (17). A distinguishing feature of this process is a check to determine if the newly entered structure also exists in the Registry file of over four million substances. If an exact duplicate is found in the CAS file, the CAS Registry Number along with the CA Index name and synonyms are returned as an update to the WL/PD Names File. Critical to the duplicate check as currently handled is that the entire structure, including the salt or solvate portion, must be identical even as to the proportion of components of a multi-component structure (e.g., $RNH_2 \cdot H_2SO_4$ does not match $RNH_2 \cdot 1/2H_2SO_4$. System modifications could remove the limitation. A profile of all WL/PD substances entered into the private WL/PD system is maintained by CAS and checked periodically for matches in the CAS Registry files. Therefore, within

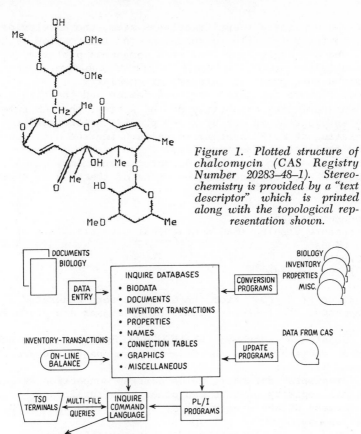

Figure 1. Plotted structure of chalcomycin (CAS Registry Number 20283–48–1). Stereochemistry is provided by a "text descriptor" which is printed along with the topological representation shown.

Figure 2. Information flow

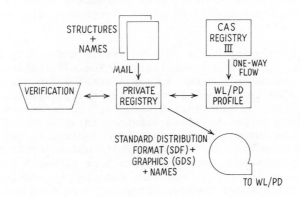

Figure 3. CAS processing

limits of the duplicate identification system, the novelty of compounds in the WL/PD file is known and the information is kept current. The CAS Registry files, with compounds published since 1965, undoubtedly contain a large fraction of the compounds characterized throughout the history of chemistry.

In addition to the full structure search and the high quality structure printing capability of the system, the CAS files and computer programs in use by WL/PD provide a straightforward way of printing at WL/PD systematic names using the many special characters familiar to chemists (4).

After keyboarding the WL/PD structures and names at CAS, the following data sets are produced by means of the private Registry for delivery to WL/PD:

- Connection tables
- Structure display (CAS's Graphical Data Structure or "GDS")
- CA names
- WL/PD names
- Editing Records

CAS Files at WL/PD. From the CAS data sets generated for WL/PD compounds, the following computer databases are produced:

- A "Substance" database, which identifies the unique substances registered and the fragment components and proportions
- A "Fragments" database, which carries connection tables, internal registry numbers and other structural data
- A "Names" database with all systematic (CAS and WL/PD) names and synonyms
- A "GDS" (for Graphical Data Structure) (12) database which holds in vector form two-dimensional representations which may be drawn by various plotting devices.

Files are shipped on a regular schedule from CAS to WL/PD, and processed, using update programs, to produce the INQUIRE databases.

Report Generation. The INQUIRE system has a very flexible set of report generation capabilities. The multi-file option allows a report to be generated from several databases at once: information such as biological screening data, chemical and physical properties, transaction history, and existence of research documents can be drawn together in one report. INQUIRE also allows the formation of output into tables, listings with headings and comments, and decoded fields which permit a compact code in the database to be expanded to more readable forms.

The chemical structure and name can also be combined with the text of the INQUIRE report. Programs have been written to format the graphical structure, the expanded character set name, and the text of the query into a single plot, which can be output on the STATUS electrostatic plotter. These plots can be copied onto either paper or card stock by XEROX reduction copying to provide both file cards and full-page reports.

Information from five different computer files are combined to provide chemical structure, name, properties, source, other compound-specific information, and biological screening results shown in Figure 4.

Routine queries can also be executed by using the "macro" feature of INQUIRE. Macros are sets of INQUIRE commands stored on the computer, and can be invoked by referring to the name of the macro. Thus, complicated multifile searches and reports can be generated with only very simple input.

System Security. The taking of confidential information out of manual office files and into remote computer systems increases the risk of disclosure regardless of who manages the system. Over a decade of successful experience in handling the National Cancer Institute's confidential files attests to the adequacy of CAS precautions. We hope that advancing technology of computer security will keep pace with the increasing use of the computer for information systems.

Hardware Configuration

The central computer at the Research Laboratories in Ann Arbor is a Varian V74, which also acts as a HASP work-station to the corporate 370/168 in Morris Plains, New Jersey. The Varian supports a full complement of peripheral equipment including line printers, disks, and tape drives (see Figure 5). The STATOS electrostatic plotter allows text and graphics to be composed simultaneously on line-printer-size pages.

The Entrex-data-entry system has replaced keypunches for most production work. When data have been entered, validated, and corrected, they may be sent on to the V74 (or through it to the 370) for further processing.

Data communications between Morris Plains and Ann Arbor is managed by a pair of Codex communications multiplexors. These devices, at either end of a single 9600 baud leased line, split its capacity between the V74 and 13 independent Trendwriter TSO terminals. Because all components are linked together at some point, the entire operation is very flexible. A user may debug a query at a TSO terminal, getting sample printouts interactively. When the report is satisfactory, a background search may be executed to process an entire file. The results, when they arrive at the V74, may be printed, held for further processing, or plotted on the STATOS if structures are included. The TSO user may continue with another query while background processing is underway.

$C_{21}H_{26}N_2O_3$ (354.45)   CN   86621
                     C267   030276
Eburnamenine-14-carboxylic acid, 14,15-dihydro-14-hydroxy-,
ester (3ā,14ā,16ā)-; Vincamine

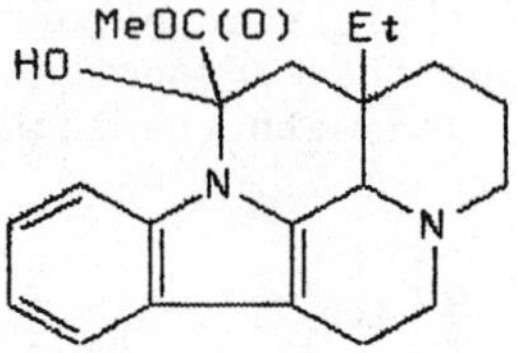

				% CHANGE
ROUTE	VEHICLE	DOSE	RATING	---------
				VOLUME
DUODENAL	4%GUM ACACIA	1.00 M/K	N	+ 11.4
DUODENAL	4%GUM ACACIA	5.00 M/K	C	- 39.2
DUODENAL	4%GUM ACACIA	10.00 M/K	A	- 57.7
PO	4%GUM ACACIA	10.00 M/K	N	+ 1.2

RATING: A = ACTIVE > OR = 50%; C = MODERATELY ACTIVE 36-49%;

COMMENTS:

Figure 4.   Biological test report

```
        (P)                               TEST:   GAH1
                                          REPORTED: 12/27/77
 methyl


              WHITE           POWDER

        MP 231-232      BP
        UNSTABLE TO:

        SOLUBILITY: ACETONE   VERY SOL.   DMA     MOD. SOL.




        TEST: SIMULATED TEST DATA
        TESTER:   DR. ROBERT JONES

        SPECIES:   RAT




FROM CONTROL
-------------       SIDE EFFECTS               TESTER'S
   TOTAL ACID                                  NOTEBOOK
-------------------------------------------------------------

    +    4.3                                   10000X000
    -   19.9                                   10000X000
    -    0.4    DEPRESSION                      10000X000
    +    3.8                                   10000X000

-------------------------------------------------------------
    D = SLIGHTLY ACTIVE 15-35%;    N = INACTIVE < 15%
```

(simulated biological data)

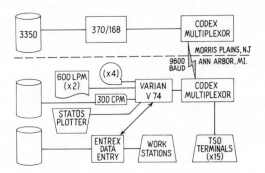

Figure 5.   Hardware configuration

Future Developments

Compatibility with the Registry System is attractive because continuing developments at CAS make the world's chemical literature more accessible. Registry Numbers obtained during processing potentially provide a direct link to abstracts. Links to data in abstracts by means of substructure search of the entire CAS structure file is a significant project recently begun at CAS (18). Conversion of systematic nomenclature to connection tables is now used as a computer edit in the CAS Registry system, and the reverse process of generating names from connection tables is being investigated (19). Standardized storage and retrieval of chemical information can lead to significant collaborative effort not only for search systems which CAS is exploring, but also for other areas such as reaction codes, computer assisted synthesis, structure elucidation, and data analysis.

Further enhancements to the hardware will probably center around some sort of interactive graphics device. This will introduce another level of hardware, since a small computer will be used to pre- and post-process the graphical data for efficiency.

Summary

Chemical Abstracts Service's (CAS) Registry III records of chemical structure are used to augment Warner-Lambert/Parke-Davis' computer files of chemical and biological data. The marriage provides a relatively low-cost system having uncommon potential. Key-to-disk entry at WL/PD of chemical properties, biological results, and other information and "on-line" weighing at WL/PD of samples are paralleled by key-to-disk entry at CAS of chemical structures and names. Computer-readable chemical structure records are returned from CAS to WL/PD. INQUIRE database management of files at WL/PD provides interactive or batch multi-file search capability and routine generation of reports. Batch processing at WL/PD is used for substructure searching and, using CAS systems installed at WL/PD, for high quality printing of structures and names. WL/PD substances are checked for duplication in the CAS Registry Structure file of over four million substances at the time of registration and periodically thereafter. CAS provides Registry Numbers and systematic names and synonyms for detected duplicates. Compatibility with CAS offers the possibility 1) for easier access to computer-readable abstracts, 2) for substructure searching of the entire CAS database, and 3) for algorithmic generation of systematic names from connection tables.

Acknowledgement

We are indebted to Mr. N. A. Farmer, Dr. D. C. Myers, Mr. T. J. Walker, Mr. C. E. Watson, and Mr. R. J. Zalac of CAS, and to Mr. F. C. Fensch, Mr. S. Fine, and Dr. A. M. Moore of WL/PD for their encouragement, advice, and many valuable discussions.

Literature Cited

1. Lynch, M. F., Harrison, J. M., Town, W. G., and Ash, J. E., "Computer Handling of Chemical Structure Information", American Elsevier, New York, N. Y., 1971.
2. Wipke, W. T., Heller, S. R., Feldmann, R. J., and Hyde, E., "Computer Representation and Manipulation of Chemical Information", Wiley, New York, N. Y., 1974.
3. Ash, J. E. and Hyde, E., "Chemical Information Systems", Wiley, New York, N. Y., 1975.
4. Dittman, P. G., Stobaugh, R. E., and Watson, C. E., "The Chemical Abstracts Service Chemical Registry System. I. General Design", _J. Chem. Inf. Comput. Sci._, (1976), 16, 111-121.
5. Schenk, H. R. and Wegmüller, F., "Substructure Search by Means of the Chemical Abstracts Service Chemical Registry II System", _J. Chem. Inf. Comput. Sci._, (1976), 16, 153-161.
6. "Agreement Links CAS, Japanese Organization", _CAS Report_, Number 6, September 1977.
7. Seals, J. V., Jr., Watson, C. E., and Wilson, G. A., "National Cancer Institute's Drug Research and Development Chemical Information System: Operation in the CAS Environment", Abstracts, 169th National Meeting of the American Chemical Society, Philadelphia, Pa., April 1975.
8. Milne, G. W. A. and Heller, S. R., "The NIH EPA Chemical Information System", Abstracts, 175th National Meeting of the American Chemical Society, Anaheim, Calif., March 1978.
9. McNulty, P. J., Garton, C. R., Buckley, S. L., Raezer, T. W., Cohen, S. G., Kippenhan, N. A., Dyott, T. M., and Gilbert, J. T., "Agricultural Chemicals Computerized Information System at Rohm and Haas Company", Abstracts, 173rd National Meeting of the American Chemical Society, New Orleans, La., March 1977.
10. Farmer, N. A. and Schehr, J. C., "A Computer-Based System for Input, Storage, and Photocomposition of Graphical Data", _Proc. Assoc. Comput. Mach._, (1974), 2, 563.
11. Dittman, P. G., Mockus, J., and Couvreur, K. M., "An Algorithmic Computer Graphics Program for Generating Chemical Structure Diagrams", _J. Chem. Inf. Comput. Sci._, (1977), 17, 186-192.

12. Blake, J. E., Farmer, N. A., and Haines, R. C., "An Inter-
     active Computer Graphics System for Processing Chemical
     Structure Diagrams", J. Chem. Inf. Comput. Sci., (1977),
     17, 223-228.
13. Stein, J. D., Jr., Delaney, F. M., Peluso, S. D., and
     Starker, L. N., "A Computer-Based Comprehensive Bio-Data
     Information Retrieval System", J. Chem. Doc., (1973), 13,
     145-152.
14. ENTREX, Inc., Burlington, Mass., a Nixdorf Company.
15. Geer, H. A., Moore, A. M., Howard, C. D., and Eady, C. E.,
     "The Parke-Davis Code for Chemical Structures", J. Chem.
     Doc., (1962), 2, 110-113.
16. Stein, J. D., Jr. and MacDougall, E. E., The Warner-Lambert
     Company, Morris Plains, New Jersey, personal communication,
     1977.
17. Dayton, D. L. and Zamora, A., "The Chemical Abstracts
     Service Chemical Registry System. V. Structure Input and
     Editing", J. Chem. Inf. Comput. Sci., (1976), 16, 219-222.
18. Farmer, N. A. and Weisgerber, D. W., "A New Strategy for
     Substructure Searching", Chemical Abstracts Service Open
     Forum, 174th National Meeting of the American Chemical
     Society, Chicago, Illinois, August 1977.
19. Vander Stouw, G. G., "Computer Programs for Editing and
     Validation of Chemical Names", J. Chem. Inf. Comput. Sci.,
     (1975), 15, 232-236.

RECEIVED August 29, 1978.

10

The NIH/EPA Chemical Information System

STEPHEN R. HELLER
Environmental Protection Agency, PM-218, Washington, DC 20460

G. W. A. MILNE
National Institutes of Health, Bethesda, MD 20014

Over the past seven years, NIH and EPA have developed a computer based Chemical Information System (CIS), which is an on-line interactive computer system that handles chemical and toxicological data (1). The CIS consists mainly of a collection of numeric (as opposed to bibliographic) data bases and software to search these data bases. The four main areas of the CIS can be grouped as follows:

1. Searchable numeric data bases
2. Structure and Nomenclature Search system (SANSS)
3. Chemical Substance Information System (CSIS)
4. Analysis and Modelling Programs

The first three areas will be described, with emphasis on the linking of areas 1 and 2.

Figure 1 shows how the four areas of the CIS are coordinated, with the Structure and Nomenclature Search System (SANSS) in the center. At present there are 25 data bases in the SANSS. These comprise the CIS Unified Data Base (UDB) and are searchable by the SANSS (2). They are shown in Figure 2. The referral aspects of the CIS represent a valuable tool for scientific and administrative work both within our respective Agencies as well as outside these Agencies, in the public and private sector, here in the USA and abroad. The referral capability of the CIS consists of a list of data bases, literature references (e.g., Merck Index) and Government Regulatory files, which can all be accessed simultaneously by consulting a single central file. All the available information concerning a substance can be located in a single operation. As the number of data bases is increased, the CIS becomes more valuable and a time-saving device in searches for chemical information. Typical questions that can be readily and inexpensively answered by this approach are:

* Has this chemical been sold as a pesticide in the USA?
* Is there a measured acute toxicity value for a particular air pollutant?
* Is information concerning a drug taken in overdose quantities and identified by gas chromatography-mass spectrometry in the Merck Index or the NIMH book on psychotropic drugs?
* Has a certain chemical been registered for sale in the USA?

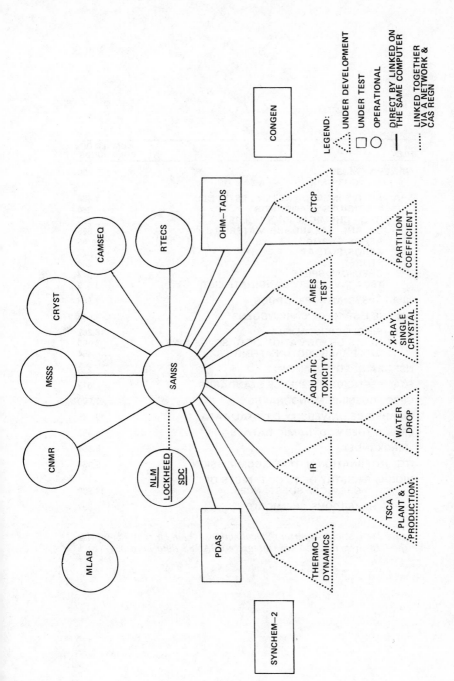

Figure 1.   The structure of the CIS with the CAS Registry number linking (CIS components—August, 1978)

FILE	NUMBER OF COMPOUNDS
NIH/EPA—MSSS	25,560
C-13 NMR	3,765
EPA—ACTIVE INGREDIENTS IN PESTICIDES	1,454
PESTICIDES STANDARDS	384
ORD—CHEMICAL PRODUCERS	375
OIL AND HAZARDOUS MATERIALS	858
AEROS/SAROAD	65
AEROS/SOTDAT	572
STORET	234
CHEMICAL SPILLS	577
TSCA INVENTORY CANDIDATE LIST	33,579
NIMH—PSYCHOTROPIC DRUGS	1,686
SRI-PHS LIST 149 OF CARCINOGENS	4,448
NBS—SINGLE CRYSTAL FILE	18,362
HEATS OF FORMATION OF GASEOUS IONS	3,169
GAS-PHASE PROTON AFFINITIES	454
NSF—RANN POLLUTANT FILE	225
FDA—PESTICIDE REFERENCE STANDARDS	613
CPSC—CHEMRIC MONOGRAPHS	1,000
CAMBRIDGE UNIVERSITY CRYSTAL DATA	10,018
EROICA THERMODYNAMIC DATA	4,492
MERCK INDEX	8,894
ITC—INTERNATIONAL TRADE COMMISSION	9,194
NIOSH—REGISTRY OF TOXIC EFFECTS OF CHEMICAL SUBSTANCES	19,908
NFPA—HAZARDOUS CHEMICALS	397

Figure 2. List of the current 25 collections which currently comprise the CIS unified data base (integrated SANSS data base 3/1/78)

Among the data bases being added to the CIS this year are
those shown in Figure 3. Over the next 2-3 years, with the con-
tinued addition of files that are either generated or used by the
Government, it is expected that the list of referral files will
grow to over 250. With the recent efforts of the four main
Federal regulatory Agencies (EPA, FDA, CPSC, OSHA) to coordinate
their various activities, such as the study and regulation of
specific chemicals, this central referral system takes on more
importance. This four-Agency group, known as the Interagency
Regulatory Liason Group (IRLG) (3), is now working to use the
Chemical Abstracts Service (CAS) Registry Number as the standard
chemical identifier for the chemicals in all the four Agencies.
An internal regulation has been proposed which will make this
mandatory. The regulation is modelled after EPA Order 2800.2,
currently the only Government regulation to mandate standardized
chemical classification (4).

Over the past four years, some 170,000 chemical names have
been submitted to CAS, under contract to EPA, to obtain the CAS
Registry Numbers for these chemicals. The result of this massive
and costly effort is the CIS Unified Data Base (UDB) of about
101,000 unique chemicals associated with the 25 files shown in
Figure 2. That there is so much overlap of the chemicals found
in these files is not surprising. It is beginning to appear that
there are relatively few chemicals which are actually studied in
any detail, and even fewer that become significant in commerce,
as, for example, drugs, food additives or pesticides. Projections
suggest that by the time the CAS registration process of some 250
files is completed, the actual size of the CIS unified Data Base
will not exceed 175,000-200,000 substances. The need then will
be to obtain as much useful and accurate information about these
substances as is necessary to protect health and environment in
the USA, as is required by the missions of our respective
Agencies. It is our hope that by defining the size or scope of
the "real" universe of chemicals, that the burden on industry will
be lessened and that future efforts will be easier to direct.
Thus, we see little immediate need to study the universe that CAS
has defined, of over some 4,000,000 chemicals found in the
literature that CAS has abstracted since 1965. Only about 12% of
these four million have appeared more than once in the CAS-
abstracted literature and probably no more than 3% are produced
and sold in anything but research quantities.

Structure and Nomenclature Search System (SANSS)

The Structure and Nomenclature Search System (SANSS), the
heart of the CIS, is based upon the work of Feldmann who developed
the original search algorithms a number of years ago (5).
Addition of a nomenclature search program, an identity search
program and a search program based on the Edgewood CIDS structure
keys (6), as well as some considerable refinement of the system

U.S. Coastguard Chemical
Properties File.

EPA IERL Non-Criteria Pollutant
Emissions.

EPA, Section 111A of the Clean
Air Act.

EPA, Office of Air Quality,
Permissible Standards,
Criteria Pollutants.

EPA, Office of Water Supply,
File of Drinking Water
Pollutants.

EPA, Pollutant Strategies Branch,
Selected Organic Air
Pollutants.

EPA, Effluent Guidelines Consent
Decree List

EPA, Section 112 of the Clean Air
Act.

EPA, ORD, Gulf Breeze, List of
Chemicals.

EPA, Carcinogen Assessment Group
List of Chemicals.

EPA, RPAR Candidates Chemical
Review Schedule List.

EPA, OTS Status Assessments.

EPA, Standing Air Monitoring
Work Group List of Non—
Criteria Pollutants.

EPA, ORD—OHEE Laboratory
Chemicals.

EPA, List of Potentially
Hazardous Chemicals from Coal
and Oil.

California OSHA List of Chemical
Contaminants.

WHO, Food and Agriculture
Organization, List of Pesticides.

EPA, IERL, Organic Chemicals
in Air.

NCI, Public List of Known
Carcinogens.

NCTR, Potential Industrial
Carcinogens and Mutagens.

EPA, IERL, List of Environmental
Carcinogens.

EPA, OPP, Pesticide Literature
Searches.

NIEHS, Laboratory Chemicals.

Toxic and Hazardous Industrial
Chemicals Safety Manual.
International Technical
Information Institute, Tokyo.

List of Teratogenic Chemicals.
Medical Information Center,
Karolinska Institute, Stockholm.

EPA, List of Hazardous Pesticides.

EPA, Mutagenicity Studies.

CITT, List of Candidates.

EPA, TSCA Section 8e, List of
Chemicals.

Figure 3. New files being added to the NIH/EPA CIS UDB in Spring,
1978

has been carried out over the last few years. The SANSS and its data base, connection tables from CAS and chemical names, has absorbed the bulk of the CIS budget.

Currently, the SANSS can be used in a number of ways. The more important methods are:

* Nomenclature Search (NPROBE)
* Ring Search (RPROBE)
* Fragment search (FPROBE)
* CIDS code search (SPROBE)
* Molecular weight search (MW)
* Molecular formula search (MF)
* Substructure search (SUBSS)
* Full structure search (IDENT)

In addition to these searching programs, there are a number of retrieval and display options available in the system. These include:

* Display of chemical structure
* Display of CAS Collective Index names
* Display of synonyms, common names and
    trade names
* Display of molecular formulas
* Display of files containing a substance
* Retrieval based upon CAS Registry Number

The following sections will be devoted to explaining the various SANSS nodules and giving examples of how they can be used. At the end of the chapter an example of the interfacing of the SANSS with the NIOSH RTECS data base of acute toxicity data (7) will be described, as an example of the direction that CIS development is taking. Since there is considerable interest on the part of the chemical industry in the implementation of TSCA, access to the bulk of the public data that EPA will be using in its work for administering TSCA should be of value. At present, development of the SANSS is being directed towards the immediate needs of EPA's Office of Toxic Substance (OTS), so that the foundation that has been built for the SANSS can be used most effectively for the implementation of TSCA.

Name - Nomenclature Search (NPROBE)

The name search, NPROBE, has been implemented as a result of requests expressed by both the SANSS user community and the CEQ-TSCA MITRE study proposal (8) for the development of a Chemical Structure and Nomenclature System which we have called the Structure and Nomenclature Search System. The software used is similar to that used in the CHEMLINE system at the National Library of Medicine (NLM) and allows for complete or partial (fragment) name search. There are an average of slightly over 3 names per chemical in CIS UDB, as opposed to slightly more than 2 names per chemical in CHEMLINE (9). The CHEMLINE file, which links primarily to the TOXLINE literature references, is made up

mostly of research chemicals, and thus is not likely to have the
multiple synonyms that are associated with commercial chemicals.
In the CIS UDB, which is comprised of files from primarily regulatory, and hence commercial, sources, there are the expected
additional names associated with materials in commerce.

To conduct a nomenclature search, the user simply enters a
chemical name or name fragment, as shown in Figure 4. The example
shown in Figure 4 is of a search for any substance in the UDB
whose name contains the fragment "DDT". From Figure 4 it can be
seen that there are 12 such substances in the UDB, of which the
first, p,p' DDT, is shown in the Figure. In addition, also shown
in this figure are all the files of the UDB which contain information on p,p' DDT, with the local file identifier numbers listed
so that one may go directly to the particular file and get the
information that is contained in that file regarding p,p' DDT. In
Figure 5, a name search for the name fragment "LSD" was performed
on the entire UDB and five examples were found. The first of
these five is shown in Figure 5, with the names of the files that
have information about LSD. Not surprisingly, the files include
the NIMH List of Psychotropic Drugs, the Merck Index and the NIOSH
acute toxicity data base, as well as the NIH/EPA Mass Spectral
Data Base and the TSCA Candidate List. There is little doubt that
the inclusion on the TSCA Candidate or "Strawman" list will be
changed once the final TSCA inventory is published, since under
present law, LSD is an illegal chemical substance. This is a useful search technique, but requires a large list of synonyms, a
correct spelling, and a knowledge of how chemical names are broken
down. For example, in searching for a cyclohexanedione, if the
file name of the substance is written as 2,5-cyclohexanedione
rather than cyclohexan-2,5-dione, a search for "dione" will not
find the chemical.

Functional Group - CIDS Key Search (SPROB)

The best way to search for functional groups or structure
features in the CIS SANSS is to use the Chemical Information Data
Systems (CIDS) keys, developed by Edgewood Arsenal. The CIDS keys,
a few of which are shown in Figure 6, are the basis of a rapid
and efficient way to search the CIS UDB for substances containing
a particular functional group or structure feature. Many of the
CIDS keys are quite specific in nature, as can be seen in Figure
6. Others, shown towards the bottom of Figure 6, are quite
generic in nature. For example, the CIDS key FG25 refers to the
presence of a nitrile or cyanide group in the molecule.

An example of a CIDS key search is given in Figure 7, where
a search is shown for all cyclohexyl (SCN49) morpholine (SCN35)
compounds in the NIOSH RTECS data base of acute toxicity. There
are only two such compounds in the data base, and the first of
these is printed out in the figure, along with its local NIOSH
RTECS identifier numbers indicated.

```
OPTION? NPROBE
FRAGMENT OR WHOLE NAME SEARCH    (F/W)   (F) ?F
SPECIFY FRAGMENT (CR TO EXIT):   DDT
FILE    1,    12 COMPOUNDS HAVING FRAGMENT: DDT
SPECIFY FRAGMENT (CR TO EXIT):    _
OPTION? SSHOW 1
HOW MANY STRUCTURES (E TO EXIT)   ?  1
TYPE E TO TERMINATE DISPLAY
STRUCTURE      1 CAS REGISTRY NUMBER 50-29-3
TSCA CANDIDATE LIST: R000-2373
CIS MASS SPECTROMETRY
CIS CARBON 13 NMR SPECTROMETRY:  50-29-3.01
EPA PESTICIDES — ACTIVE INGREDIENTS:  29201
EPA OHM/TADS:  72T16510
CAMBRIDGE XRAY CRYSTAL:  50-29-3.01
MERCK INDEX
EPA PESTICIDES — ANALYTICAL REF. STNDS.:  1880, 1920
EPA STORET:  39317, 39373, 39371, 39374, 39372, 39370, 39359, 39375
    39376, 39378, 39290, 39358, 39377, 39302, 39303, 39304, 39300, 39301

EPA CHEMICAL SPILLS
CPCS CHEMRIC
FDA/EPA PESTICIDES REF. STANDARDS:  200
U.S. INTERNATIONAL TRADE COMMISSION
NBS XRAY CRYSTAL:  50-29-3.01
NSF CHEMICALS LIST:  138
PHS-149 CARCINOGENS:  A0240
NIOSH RTECS:  KJ33250
```

C14H9C15

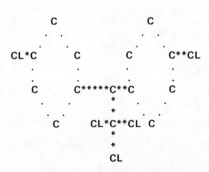

Benzene, 1, 1, — (2, 2, 2-trichloroethylidene) bis [4-chloro—   (9CI)
Ethane, 1, 1, 1-trichloro-2, 2-bis (p-chlorophenyl)—   (8CI)
.alpha., .alpha. —Bis (p-chlorophenyl)—.beta., .beta., .beta. -trichlorethane
p, p' -Dichlorodiphenyltrichloroethane
p, p' -DDT

Figure 4.   NPROBE name search for name fragment "DDT"

```
OPTION? NPROBE
FRAGMENT OR WHOLE NAME SEARCH  (F/W)  (F) ?F
SPECIFY FRAGMENT (CR TO EXIT):        LSD
FILE   5,      5 COMPOUNDS HAVING FRAGMENT:  , LSD
SPECIFY FRAGMENT (CR TO EXIT):
OPTION? SSHOW 5
HOW MANY STRUCTURES (E TO EXIT) ?  1
TYPE E TO TERMINATE DISPLAY
STRUCTURE     1 CAS REGISTRY NUMBER 50-37-3
TSCA CANDIDATE LIST:  R000-3157
CIS MASS SPECTROMETRY
MERCK INDEX
NIMH PSYCHOTROPIC DRUGS: 273
NIOSH RTECS:  KE42000,KE41000,KE43750
```

C20H25N30

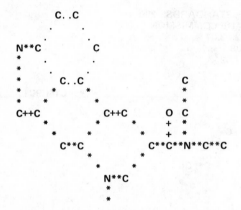

Ergoline-8-carboxamide, 9, 10-didehydro-N, N-diethyl-6-methyl–, (8.beta.) –
  (9CI)
Ergoline-8 .beta. -carboxamide, 9, 10-didehydro-N, N-diethyl-6-methyl– (8CI)
(+) –LSD
D–LYsergic acid diethylamide
D–Lysergic acid N, N-diethylamide

Figure 5.   NPROBE name search for LSD

Key	Structure
SCN 1	
SCN 35	
SCN 48	
SCN 49	
FG 25	
FG 39	
FG 87	
FG 116	
FG 219	

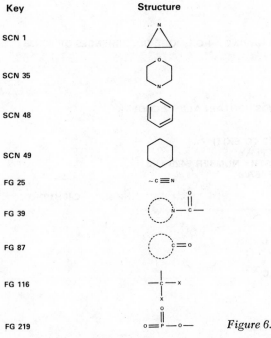

Figure 6.   Sample CIDS key codes

```
OPTION? SPROBE
SPECIFY STRUCTURAL FEATURE CODE AND PERMISSIBLE MULTIPLICITY LIMITS
NEXT SFC = SCN49
FOUND    428 COMPOUNDS HAVING   1 OR MORE OCCURRENCES OF SCN49


. . . . . . . . . . . . . .

NEXT SFC = SCN35
FOUND    277 COMPOUNDS HAVING   1 OR MORE OCCURRENCES OF SCN35


. . . . . . . . . . . . .

NEXT SFC = _
FILE = 11,   ‾ 2 COMPOUNDS CONTAIN ALL   2 CODES

OPTION? SSHOW 11
HOW MANY STRUCTURES (E TO EXIT)  ?  1
TYPE E TO TERMINATE DISPLAY
STRUCTURE    1 CAS REGISTRY NUMBER 6425-41-8
NIOSH RTECS: QE06400,QE06700
```

$C_{10}H_{19}NO$

Morpholine, 4-cyclohexyl— (8CI9CI)
Cyclohexylmorpholine
N-Cyclohexylmorpholine
4-Cyclohexylmorpholine

Figure 7.   CIDS key search for cyclohexyl morpholine compounds

Molecular Weight (MW) and Formula (MF) Search

In addition to searching for a particular functional group using the CIDS keys as shown above, it is possible to search for a compound, or a group of compounds, using molecular weight. The molecular weight search, shown in Figure 8, allows for either a specific molecular weight, or, as is indicated in the figure, a range of molecular weights. In the particular example shown in Figure 8, the Merck Index is being searched for all occurrences of compounds with a molecular weight between 368 and 380. There are 167 such substances as can be seen in the top part of Figure 8. This is too large a number and so it was decided to try to narrow or filter the search down to a smaller number using a molecular formula search. In this case what was really sought were all compounds which have two oxygen atoms and a molecular weight between 368 and 380. In Figure 8 a search for this partial formula (O2) is shown, and this is followed by a Boolean AND logic operation (INTERsect) between the file of 167 compounds with the correct molecular weight range and the file of 1484 having the correct partial formula. The result of this AND operation is a file containing the 16 compounds in the Merck Index which have a molecular weight between 368 and 380 as well as exactly two oxygen atoms in the molecule. At the bottom of Figure 8, the first of the 16 answers is printed out. This compound, with a molecular formula of C21.H23.ClF.N.O2 and a molecular weight of 375, is Haloperidol, which is a drug used as a sedative and tranquilizer.

In the event that there is no interest in chlorinated compounds, even though they may meet the molecular weight and molecular formula criteria, a further molecular formula search may be conducted, as shown in Figure 9, for compounds with 1-4 chlorine atoms. From Figure 9, it can be seen that there are 986 compounds with 1-4 chlorine atoms in the Merck Index file. Since the requirement was for compounds that did not contain this halogen atom, a Boolean NOT operation between the 986 chlorine-containing compounds and the 16 compounds previously found is performed, as seen in the center of Figure 9. This results in the removal of three of the sixteen substances, and of the remaining thirteen, the first one, Androsta-3,5-dien-17-ol, 3-(cyclopentyloxy)-17-methyl-, (17.beta.), is printed out and shown at the bottom of Figure 9. This, of course, like the other twelve in the file, does not contain the chlorine that was present in three of the answers to the first search shown in Figure 8. The ability to interact and impose various limitations and filters on searching is a very powerful capability of the SANSS.

```
OPTION? MW
TYPE MW OR RANGE, CR TO EXIT
USER: 368-380
FILE =  4,     167 COMPOUNDS WITH MW     368-380
OPTION? MF
CR TO EXIT, COMPLETE (C), PARTIAL (P), OR RANGED (R) MF?
USER: P
THE NUMBER OF ATOM TYPES IS: 1
ENTER ATOM, FOLLOWED BY COUNT FOR EACH TYPE, E.G. C6.
TYPE 1 IS: 02
FILE =  5,     1484 COMPOUNDS HAVING PARTIAL MF: 02


•••••••••••••

CR TO EXIT, COMPLETE (C), PARTIAL (P), OR RANGED (R) MF?
USER: ___
OPTION? INTER 4 5
FILE =  6       RESULTING REFERENCES =        16
SOURCE FILES WERE:    4     5
OPTION? SSHOW 6
HOW MANY STRUCTURES (E TO EXIT) ? 1
TYPE E TO TERMINATE DISPLAY
STRUCTURE     1 CAS REGISTRY NUMBER 52-86-8
MERCK INDEX
```

C21H23ClFNO2

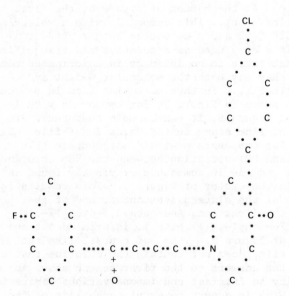

1-Butanone, 4-[4-(4-chlorophenyl) -4-hydroxy-1-piperidinyl]-1-(4-fluoro
    phenyl)- (9Cl)

Figure 8.   Molecular-weight range search

```
OPTION? MF
CR TO EXIT, COMPLETE (C), PARTIAL (P), OR RANGED (R) MF?
USER: R
THE NUMBER OF ATOM TYPES IS: 1
ENTER ATOM, FOLLOWED BY RANGE FOR EACH TYPE, E.G. C6,12.
TYPE 1 IS: CL1,4
FILE =  7,    986 COMPOUNDS HAVING PARTIAL MF IN RANGE: CL1-4


OPTION? NOT 6 7
FILE =  8     RESULTING REFERENCES =    13
SOURCE FILES WERE:    6    7
OPTION? SSHOW 8
HOW MANY STRUCTURES (E TO EXIT) ? 5
TYPE E TO TERMINATE DISPLAY
STRUCTURE     1 CAS REGISTRY NUMBER 67-81-2
MERCK INDEX
```

C25H38O2

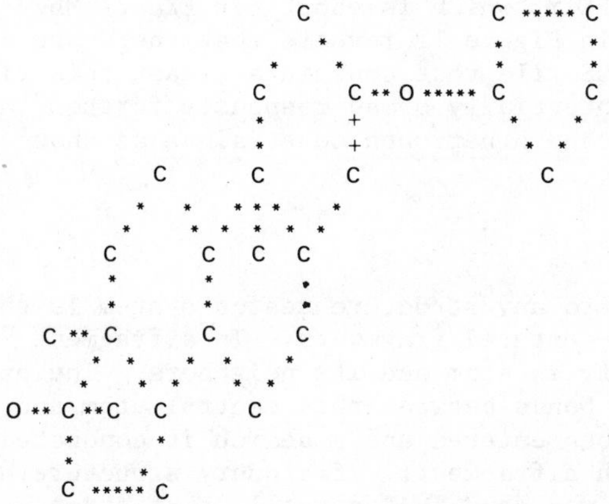

Androsta-3,5-dien-17-ol, 3-(cyclopentyloxy)-17-methyl-, (17.beta.)- (9
          Cl)

Figure 9.   Sample of combination searches of MF, MW with NOT logic

Nucleus - Ring Search (RPROBE)

    One of the features of the CIS SANSS that has made the system
useful is the structure of the file with respect to ring systems.
The SANSS has a hierarchical file structure that allows for rapid
and inexpensive searching for specific rings or ring systems.   In
Figure 10, a list of some of the commands used to generate structures are given. To show how the SANSS works and how one can use
the various query modules, the remainder of the chapter will be
devoted to searching through the NIOSH TTECS data base for chemicals having an aromatic ring, substituted on ortho carbons with
chlorine and bromine respectively. The first thing that must be
done in order to perform such a search is to build the 'query'
structure that is to be sought. This is done with the first few
commands shown in Figure 11. The query structure in Figure 11 is
a chloro bromo (ortho) substituted benzene ring, but the ring
probe search will be conducted for any ortho disubstituted aromatic ring, since it does not take into account the nature of the
substituents. Also, since other substituents on the benzene ring
will be permitted, it is necessary to reset the substituent search
level from 'EXACT' (only two substituents and these must be ortho)
to 'IMBED' (there must be two ortho substituents at a minimum).
The command to do this is EXIM, which is short for EXact/IMbed
switch. The search shown in Figure 11 reveals that there are 2715
compounds in the NIOSH RTECS file that contain at least this ring
pattern. To filter such potentially broad responses further, one
can use CIDS keys searches and other such constraints as shown
below.

Fragment Search (FPROBE)

    One feature necessary to any structure search system is the
ability to search for atom-centered fragments. In a fragment
search the user must specify an atom and its neighbors. The exact
(or generic) nature of the bonds between this central atom and
each of its neighbors is then entered and a search is conducted
for all occurrences of such a fragment. If a query structure has
already been generated, as was done in Figure 11, that structure
can be used by the SANSS program to generate and search for fragments. There are usually a number of atoms in a query structure
that can be considered as central to a fragment. Hence, a request
for a fragment probe of the substructure shown in Figure 11 would
lead to searches for six fragments, four of which would be the
same (i.e. atom centered fragments about atoms 3, 4, 5 and 6 are
all the same, representing a carbon atom in an aromatic ring
attached to two other aromatic carbon atoms in the ring and a
hydrogen). Such fragments are not very specific, and so it is
best to identify the atom centered fragment for which one wishes
to search. In Figure 12, atom number 1 is selected and a search
for all occurrences of a chlorine atom on an aromatic ring is

COMMAND	EFFECT
AATOM n1 m1	Insert an atom between atom n1 and atom m1.
ABOND n1 m1	Insert a bond between n1 and m1.
ABRAN l1 at n1	Add a branch of length l1 at atom n1.
ALINK n1 l1 m1	Insert a chain of length l1 between n1 and m1.
ALTBD n1 m1	Define alternate bonds in the smallest ring containing n1 and m1 as aromatic bonds.
ARING n1 m1 l1	Create a ring of l1 atoms between n1 and m1.
CHAIN l	Create a chain of l atoms.
CLEAR	Erase the existing query structure.
CRING n1 l1	Create a ring of l1 atoms including atom n1.
DATOM n1	Delete atom n1.
DBOND n1 m1	Delete the bond joining nl and m1.
MORGA	Renumber the query structure by the Morgan algorithm.
NUC 66	Create a structure of two fused six-membered rings.
REG	Retrieve the structure corresponding to a specific registry number.
REST	Negate the effect of the previous command.
RING l	Create a ring of l atoms.
SATOM n1	Define the elemental nature of atom n1.
SBOND n1 m1	Define the nature of the bond joining n1 and m1.
SPIRO n1 l1	Create a spiro-attached ring of (l1 +1) atoms at n1.
WISBD n1 m1	Define alternate bonds in the smallest ring containing n 1 and m1 as double bonds.

Figure 10. Commands used to generate structures for search-ing

```
ENTER NEW SELECTION  (H FOR HELP):  32

COLLECTION SELECTED:  32
OPTION?
OPTION?  RING
OPTION?  ABRAN  1 AT 1  1 AT 2
OPTION?  SATOM 7
SPECIFY ELEMENT SYMBOL = CL
OPTION?  SATOM 8
SPECIFY ELEMENT SYMBOL = BR
OPTION?  ALTBD  1  2
OPTION?  D
         3 . . 4
               .              .
8BR2                 5
               .              .

           1 . . 6
           ?
           ?
           7CL

OPTION?  EXIM
SPECIFY SEARCH LEVELS TO BE CHANGED
LEVELS = 4
OPTION?  RPROBE
      C??C
    ?      ?
  ?          ?
C              C??
  ?          ?
    ?      ?
      C??C
        ?.
          ?
```

                    CONDITIONS OF SEARCH
          CHARACTERISTICS TO BE MATCHED   TYPE OF MATCH
TYPE OF RING OR NUCLEUS   EXACT
NO HETEROATOMS   EXACT
SUBSTITUENTS AT   1   2   IMBED
THIS RING/NUCLEUS OCCURS IN 2715 COMPOUNDS

FILE = 1, 2715 COMPOUNDS CONTAIN THIS RING/NUCLEUS

Figure 11.   A ring-probe (RPROBE) search for a disubstituted
benzene

performed. The result of this search is a file containing all
1618 compounds in the NIOSH RTECS file that contain this particular structure fragment.

After the fragment search is conducted for the chloro
aromatic fragment, a similar search is performed on the fragment
centered about atom 2, which contains a bromo substituent. This
fragment probe (FPROBE) search, shown in Figure 13, results in
229 occurrences of this fragment in compounds in the NIOSH RTECS
data base.

Substructure Search (SUBSS)

The Substructure Search option is an atom-by-atom, bond-by-
bond comparison between connection tables in the data base and the
connection tables corresponding to the query structure. This time
consuming, sequential search is quite costly and so the ring
probe, fragment probe, and other search techniques described above
are used as screens to speed up the process and reduce the cost.
Following the three separate searches done in Figures 11-13, the
next step is to see which compounds in the NIOSH RTECS data base
contain occurrences of all three. This is done by a simple
Boolean AND logic combination of the three lists of Registry
Numbers generated by the searches in these Figures. The intersection of the lists, performed by the INTER command as shown in
Figure 14, results in 12 compounds meeting the criteria of all
three searches. However, not necessarily all of the 12 answers
are precisely what is wanted. This is because the three searches
in Figures 11-13 are for "pieces" of the structure sought but the
searches do not require these pieces to be in the same juxta-
position as in the query structure. That is, the three requirements comprise a necessary, but not sufficient condition for an
answer to the original question. To secure an exact answer as to
how many (if any) of these 12 compounds meet the exact query
structure, it is necessary to perform a true substructure search
(SUBSS) as is shown in Figure 14. The result of the use of
SUBSS shows that only 7 of 12 "answers" from the intersection of
the three searches do have the bromine and chlorine ortho to one
another on the benzene ring. Of the 7 answers, one is shown in
Figure 15. As it turns out from inspection of all 12 prior
answers (not shown here), the other compounds retrieved are meta
substituted chloro bromo aromatic compounds.

Complete Structure Search (IDENT)

The final SANSS module to be described in this chapter is
the search for a total or full structure, rather than a substructure. This module was designed primarily for the purpose of
searching for and reporting specific chemicals as part of the
TSCA inventory reporting procedures. The full structure search,
called IDENT (for IDENTity), has and will continue to have

OPTION? <u>FPROBE 1</u>

TYPE E TO EXIT FROM ALL SEARCHES,
T TO PROCEED TO NEXT FRAGMENT SEARCH

FRAGMENT:

        7CL????1C.....6C
          .
          .
          .
          2C

REQUIRED OCCURRENCES FOR HIT : 1
THIS FRAGMENT OCCURS IN   1618 COMPOUNDS

FILE = 2,   1618 COMPOUNDS CONTAIN THIS FRAGMENT

*Figure 12. A fragment probe (FPROBE) for a chlorine
atom attached to an aromatic carbon atom*

OPTION? <u>FPROBE 2</u>

TYPE E TO EXIT FROM ALL SEARCHES,
T TO PROCEED TO NEXT FRAGMENT SEARCH

FRAGMENT:

        8BR????2C.....1C
          .
          .
          .
          3C

REQUIRED OCCURRENCES FOR HIT : 1
THIS FRAGMENT OCCURS IN   229 COMPOUNDS

FILE = 3,   229 COMPOUNDS CONTAIN THIS FRAGMENT

*Figure 13. A fragment probe (FPROBE) for a bro-
mine atom attached to an aromatic carbon atom*

OPTION? <u>INTER 1 2 3</u>
FILE = 4,   RESULTING REFERENCES =   12
SOURCE FILES WERE: 1   2   3
OPTION? SUBSSS 4
DOING SUB-STRUCTURE SEARCH
TYPE E TO EXIT

FILE ITEM   10 STRUCTURE BEING SEARCHED   21609905
HITS SO FAR   6

FILE = 5,   SUCCESSFUL SUB STRUCTURES =   7

*Figure 14. Intersection and substructure search of files de-
rived in Figures 11–13*

STRUCTURE   7 CAS REGISTRY NUMBER 4824-78-6
NIOSH RTECS: TE70000

C10H12BrC12O3PS

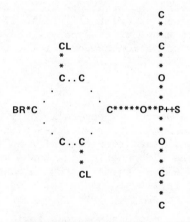

Phosphorothioic acid, O—(4-bromo-2, 5-dichlorophenyl) O, O—diethyl ester
    (8CI9CI)
Bromophos-ethyl
Ethyl bromophos
Filariol 60
Nexagan G

Figure 15.   One of seven substructure search hits

specific application to TSCA activities. For example, after the
final "grandfather" inventory required under section 8 of the
Act is published and made available, via the CIS, as well as by
other means, it will be necessary for potential vendors of a
chemical to determine if the chemical they wish to see or manu-
facture is in the Inventory and can thus be produced and marketed
without extensive pre-manufacturing testing. Use of the IDENT
search will quickly reveal if the chemical is in the TSCA inventory. Of course, one can use the name search capabilities,
but there is no guarantee that the name used by the manufacturer
will be in the list of synonyms associated with the inventory.
The structure shown in Figure 16 was generated using the standard
SANSS structure generation commands, such as those listed in
Figure 10. The IDENT search was then invoked and after being told
that the structure had the normal number of hydrogen atoms, consistent with normal valence, it found the structure in the CIS
UDB. The structure was then printed out, with all the local file
identifier information, as well as a number of synonyms, one of
which is the TSCA Clerical Code Designation number for the substance.

SANSS-Data Base Interfaces

A structure or a nomenclature search is generally only a
means to an end. The end is often some data associated with the
structures found. In order to facilitate retrieval of such information, an interface between the CIS numeric data bases and the
SANSS has been constructed. This allows for a search through the
UDB followed by a data search (or retrieval) and permits one to
answer such queries as:

* Do any ortho bromo-chloro aromatic compounds have a toxicity
  greater than 1.0 mg./kg?

In the example shown in Figure 17, the first three answers
from the previous search are used to retrieve the toxicity data
associated with these compounds. The automatic interface between
the systems is invoked by the command TSHOW and then the previous
file of 7 CAS Registry Numbers, generated by SUBSSS, are specified, with only the first three being printed out upon request.

Summary

The NIH/EPA CIS has developed to the point where complex
questions can be readily answered. The ability to manipulate
structure and numeric data and establish correlations between the
two should be of considerable value to the EPA in its work under
the Toxic Substances Control Act, as well as to scientists in
general. The value of the SANSS linked to CNMR data has been
recently shown (10), and no doubt other structure-data studies

OPTION? D

```
   10CL  70
    ?    +
    ?    +
8CL3??1??2P?5O?11
    ?  ?  ?
    ?  ?  ?
   9CL4O 60
       ?
       ?
      12
```

OPTION? IDENT

TOTAL PROTON COUNT FOR THIS STRUCTURE IS
        (P FOR PROGRAM ESTIMATE) : P
TOTAL PROTON COUNT BASED UPON NORMAL CONDITIONS IS 8
ARE THERE ANY ABNORMAL VALENCE OR CHARGE CONDITIONS WHICH
        WOULD AFFECT THIS COUNT (Y/N) ? N
PROTON COUNT FOR NODE 2 (D TO DISPLAY STRUCTURE) ?
FILE 10, THIS STRUCTURE IS CONTAINED IN 1 COMPOUNDS.
OPTION? SSHOW 10
STRUCTURE   1 CAS REGISTRY NUMBER 52-68-6
TSCA CANDIDATE LIST: R001-5032
EPA PESTICIDES − ACTIVE INGREDIENTS: 57901
EPA OHM/TADS: 72T16519

CAMBRIDGE XRAY CRYSTAL: 52-68-6.01
MERCK INDEX
EPA PESTICIDES − ANALYTICAL REF. STNDS.: 6780
EPA CHEMICAL SPILLS
FDA/EPA PESTICIDES REF. STANDARDS: 48
PHS-149 CARCINOGENS: C0147
NIOSH RTECS: TA07000

                                                C4H8C13O4P

```
   CL    O
    *    +
    *    +
CL*C**C**P**O**C
    *  *  *
    *  *  *
   CL  O  O
          *
          *
          C
```

Phosphonic acid, (2, 2, 2-trichloro-1-hydroxyethyl)−, dimethyl ester (8C
    I9CI)
Agroforotox
Anthon
Bayer L 13/59
Chlorofos

Figure 16.   Example of IDENT search for a complete molecule

DATA BASE IS NOW RTECS

OPTION? RETRIEVE
NUMBERING SYSTEM? CAS
SOURCE? FILE 5

THERE WERE   7 NUMBERS FOUND IN FILE 5
DISPLAY HOW MANY? (TYPE E TO EXIT) 3
CAS NUMBER =   2104963 NIOSH NUMBER = TE71750
  ORL-RAT LD50: 1600 MG/K TFX: TXAPA9 14,515,69
  SKN-RBT LD50:   720 MG/K TFX: GUCHAZ 6,54,73
  UNK-MAM LD50: 2000 MG/K TFX: 30ZDA9 −,335,71
  Phosphorothioic acid, O−(4-bromo-2, 5-dichlorophenyl) O, O-dim ethyl
  ester (8CI9CI)
  C8H8BrCl2O3PS

CAS NUMBER =   2720174 NIOSH NUMBER = TB01850
  ORL-RAT LD50:   35 MG/K TFX: ARSIM* 20,6,66
  ORL-MUS LD50:   77 MG/K TFX: ARSIM* 20,6,66
  Phosphonothioic acid, ethyl−, O−(4-bromo-2, 5-dichlorophenyl)
   O-ethyl ester (8CI9CI)
  C10H12BrCl2O2PS

CAS NUMBER =   2720185 NIOSH NUMBER = TB10700
  ORL-RAT LD50:   73 MG/K TFX: ARSIM* 20,6,66
  Phosphonothioic acid, methyl−, O−(4-bromo-2, 5-dichlorophenyl
   O−(1-methylethyl) ester (9CI)
  C10H12BrCl2O2PS

Figure 17.   Example of NIOSH RTECS toxicity data retrieval will be undertaken now that the necessary groundwork has been laid.

Acknowledgements

The authors wish to thank the following for their help and cooperation in developing the CIS SANSS: R. J. Feldmann, W. Greenstreet, M. Yaguda, M. Bracken, A. Fein, G. Marquart, and J. Miller.

Literature Cited

1. Heller, S.R., Milne, G.W.A., and Feldmann, R.J., Science, (1977), 195, 253.
2. Feldmann, R.J., Milne, G.W.A., Heller, S.R., Fein, A., Miller, J.A., and Koch, B., J. Chem. Info. and Comp. Sci., (1977), 17, 157.
3. The Interagency Regulatory Liason Group (IRLG) was established 2 August, 1977 by the following four Agencies: EPA, FDA, OSHA and CPSC.
4. EPA Order #2800.2, issued 27 May, 1975.
5. Feldmann, R.J., and Heller, S.R., J. Chem. Doc., (1972), 12, 48.
6. CIDS Structure Feature Key Code Manual is available from CIS Project, Chemistry Department, Brookhaven National Laboratory, Upton, Long Island, New York 11973.
7. NIOSH, Registry of Toxic Effects of Chemical Substances (RTECS), 1977. Available from the US Government Printing Office, GPO Order Number 017-033-0027101; $17.50 per copy USA: $21.88 per copy non-USA.
8. Bracken, M., Dorigan, J., Hushon, J., and Overbey, II, J., MITRE Reprt MIR-7558 to CEQ, June 1977. Two volumes entitled "Chemical Substances Information Network (CSIN)".
9. NLM Fact Sheet for the Toxicology Information Program, January 1978.
10. Milne, G.W.A., Zupan, J., Heller, S.R., and Miller, J.A., Anal. Chim. Acta, In press (1978).

RECEIVED August 29, 1978.

11

An Integrated System for Conducting Chemical and Biological Searches

T. M. DYOTT, A. M. EDLING, C. R. GARTON, W. O. JOHNSON,
P. J. McNULTY, and G. S. ZANDER

Rohm and Haas Company, Norristown Road, Spring House, PA 19477

Over the past seven years we at Rohm and Haas Company have been developing a computerized chemical and biological information system called ACCIS (Agricultural Chemicals Computerized Information System)(1). In this paper we will describe the chemical and biological search capabilities which we have built into ACCIS.

ACCIS Design Criteria

ACCIS was developed in order to:

1. Accomodate the growing amount of data which resulted from expanding biological screening programs.

2. Facilitate communication of screening results to researchers, administrators, and outside collaborators.

3. Reduce the time our biologists spent transcribing, extracting, and reporting screening results.

4. Enhance the value of the stored screening results by making them readily available.

To meet these objectives we decided that the system must:

1. contain not only the biological screening results, but also the chemical structures, reference data, and pertinent chemical data, e.g., solubility and purity information.

2. produce a variety of current awareness reports on standard 8 1/2 x 11 paper, or 3 x 5 or 5 x 8 cards, and that those reports should contain high quality structural diagrams whenever appropriate.

3. provide a convenient mechanism for conducting a
    wide variety of chemical and/or biological searches.

System Organization

ACCIS is thoroughly integrated into a everyday operation
of our screening programs. The flow of information into ACCIS
is diagrammed in Figure 1. When our chemists synthesize a
compound they complete a compound submittal form, giving the
empirical formula, structural diagram, chemical name, chemist's
name, notebook reference, department, date, various physical
properties, screening priorities, and any special instructions.
The chemist then takes the submittal form and the sample itself
to the Screening Information Center. There the information is
reviewed and entered into the system via a chemical typewriter
(a modified IBM MCST). Sub-samples are then weighed out and
sent to the appropriate screening area(s) along with a computer-
produced transmittal sheet which provides the biologists with
the structural diagram, useful physical property information,
and any special instructions. The biologists then screen the
compound, recording their findings on 2-part carbonless forms.
They keep the first copy as a legal record, while the second
copy is returned to the information center where the data are
keypunched and read into the system. Whenever data are entered,
various current awareness reports are automatically generated
which keep the chemists, biologists, and their management in-
formed and allow them to maintain hardcopy files. A typical
ACCIS report, the herbicide current awareness report, is shown
in Figure 2. (The organism names have been replaced by the
letters B-L for confidentiality reasons.) AM and AD are average
control data for all monocot and all dicot species, respectively.

The number of screening programs fluctuates as new programs
are initiated and old ones are terminated, but is generally in
the range of 8-12. Each screen may in turn include anywhere
from 1 to 15 different organisms, treated under various conditions and dosages. This variability makes it essential that
the biologists in each area work closely with the information
specialist to design both their data collection forms and the
various reports they require. Our emphasis is on meeting the
researcher's needs rather than simplifying the programming.
As a result ACCIS:

1. is a highly customized system.

2. consists of well over 100 programs, totaling
    approximately 250,000 lines of code.

3. enjoys extremely strong user support.

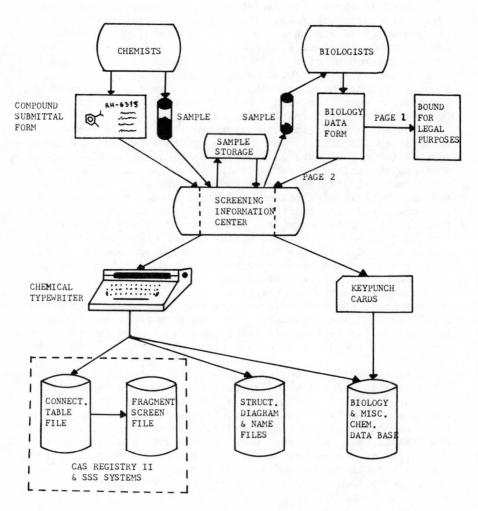

Figure 1. Flow of information into ACCIS

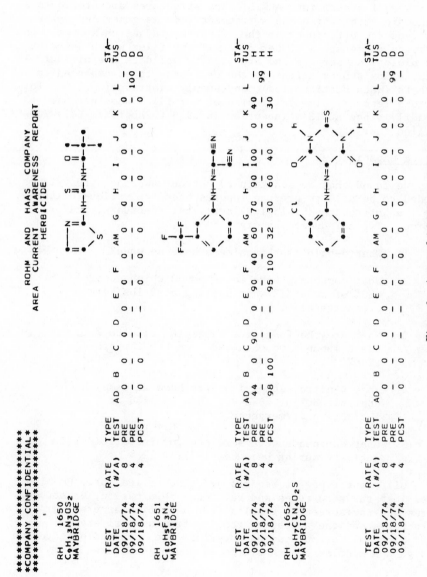

Figure 2. A typical ACCIS report format

The chemical and biological information in ACCIS is stored in a number of computer files. The biological, miscellaneous chemical, and reference information is stored in an IMS data base. The structural diagram, as entered on the chemical typewriter, and the chemical name are stored in standard variable record length files. In order to store the chemical structures in a machine intelligible, and therefore searchable, manner we incorporated the Chemical Abstracts Service (CAS) Registry II system into ACCIS. The structures are stored in a connection table file and a fragment file is generated which improves the efficiency of the substructure search system. In addition there are a number of auxiliary files which describe the biological screens and are used to validate the biological data, allow abbreviations in the data base to be expanded in reports (data dictionaries), and supply distribution lists for various reports. The total size of our files has increased steadily since ACCIS's inception in 1973 to approximately 200 million characters.

Search Capabilities

We found that in addition to current awareness reports we needed to be able to produce reports based on various criteria, e.g., substructure, biological activity, test date, and/or source. Typical questions might be:

1. What 5-halo isothiazalones have we made?

2. What compounds have we screened which control >80% of weed XYZ when applied at 2 lbs/acre preemergence?

3. What are the fungicide screening results for the compounds we obtained from KLM corporation?

4. What 4-nitro diphenyl-ethers have we made which control >80% of weed RST when applied at 4 lbs/acre postemergence?

5. What compounds were screened for insecticidal activity during December 1977?

Different types of reports are also called for. We might need just the structures and reference information, or structures and the screening results from a particular area, or structures and the screening results from several areas.

Since all of our common questions are compound oriented we designed a modular search system as shown in Figure 3.

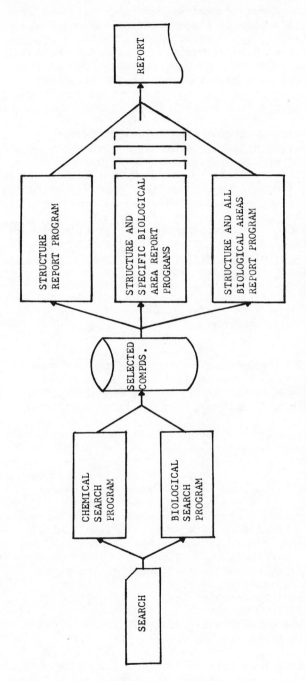

Figure 3. Flow diagram of modular ACCIS search system

A suitable chemical search program for CAS Registry II files had already been developed by CAS, while the various report programs are modified versions of current awareness report programs we have previously developed. The only major new program we needed was one for searching the biological and reference information contained in the IMS data base.

Biological Search

The biological data we need to search is contained in an IMS data base, which has a hierarchical structure, as shown in Figure 4. This hierarchical structure allows you to have any number of test areas within a compound, any number of test dates within a test area, any number of test types within a test date, etc. (There is of course more detailed information within each segment of the data base than we have depicted.)

We developed a search program which provides a very general search capability. It allows us to qualify the search or any piece (or pieces) of information in the data base and has considerable Boolean logic capabilities. For example, if we were interested in compounds within the range RH-60000 to RH-80000 which were active against fungus ABC or DEF, but did not injure crop XYZ at a rate of 4 lbs/acre, we would encode the question as:

	Comments
(RH>60000*RH<80000)	* is a Boolean AND
(AREA=F)	F for fungicide
(ORGANISM=ABC+ORGANISM=DEF)	+ is a Boolean OR
(ACTIVITY=A+ACTIVITY=B)	fungicide activities are stored as letters
(AREA=H)	H for herbicide
(DOSE=4.00)	
(ORGANISM=XYZ)	
(ACTIVITY=0)	herbicide activities are stored as numbers

The biological search system operates in a batch mode. While it would be nice to have an interactive biological search system we are confronted with a number of problems, including:

1. our reports contain structural diagrams, requiring either a special line printer or a graphics terminal.

2. most of our files are sequential rather than random access.

We do not feel that it is critical for our system to be interactive because our batch system turn-around is usually on

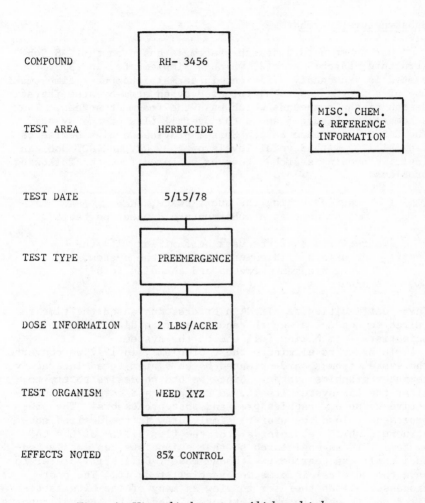

COMPOUND — RH– 3456

TEST AREA — HERBICIDE — MISC. CHEM. & REFERENCE INFORMATION

TEST DATE — 5/15/78

TEST TYPE — PREEMERGENCE

DOSE INFORMATION — 2 LBS/ACRE

TEST ORGANISM — WEED XYZ

EFFECTS NOTED — 85% CONTROL

Figure 4.   Hierarchical structure of biology data base

the order of 30-60 minutes. We are, however, planning to develop a conversational interactive front-end program which will allow us to conveniently enter a search from a keyboard terminal in our own laboratory and will automatically initiate the batch search program.

Substructure Search

Our substructure search system is based on the CAS Substructure Search System(2) which we implemented in 1975. This system is reasonably efficient in terms of computer time, making use of a fragment screen, followed by an atom-by-atom iterative search of the compounds which pass the fragment screen. The schematic in Figure 5 shows the overall flow of the system. The CAS Substructure Search System provides a high degree of flexibility, with several levels of AND, OR, and NOT Boolean logic. Encoding searches however suffered from the following problems:

1. required coding the query twice, once as fragments, once as a substructure connection table.

2. required that the user be familiar with the details and idiosynchrasies of the system, e.g., fragment screens and encoding techniques.

These difficulties resulted in errors, increased the time required to conduct a search, and, therefore discouraged our scientists from making full use of the system.

In order to eliminate these problems, in 1977 we converted the system from a card-oriented batch system to an interactive computer graphics system. Since we had no desire to try to alter the CAS system itself, we designed and developed interactive computer graphics pre- and post-processors. The pre-processor allows the user to simply draw in the desired substructure while retaining all of the flexibility of the CAS system. It was engineered so that scientists can easily conduct their own searches, although most (~80%) still prefer to have the chemical information specialist do it. The post-processor allows the user to view as many of the hits as desired and to specify which structural and/or biological reports should be prepared. The specified reports are then printed by a background batch job (automatically initiated by the post-processor) and are usually available within 30-60 minutes. The flow of this system is depicted in Figure 6.

Using the system is very straight forward. If we want to search for

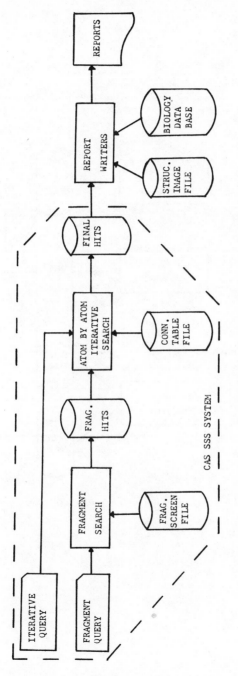

Figure 5. Flow diagram of CAS substructure search system

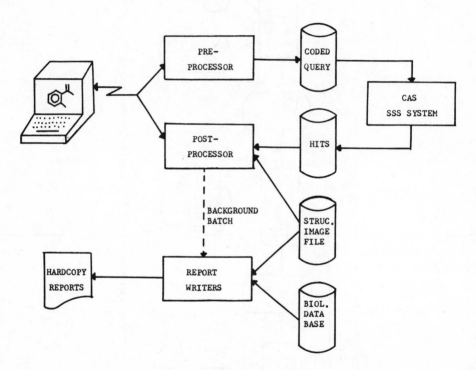

Figure 6.   Flow diagram of interactive substructure search system

we merely draw the following on the graphics terminal, via a joystick, cursor, light buttons.

A1 to A10 stand for alternatives. @1 means the atom can only have 1 non-hydrogen attachment, @2 two non-hydrogen attachments, etc.

With this system a typical search can be completed in 10-20 minutes, which can be broken down as follows:

1.	establishing computer link	1 minute
2.	enter substructure	2-4
3.	search	3-5
4.	view structures hit	3-10
5.	specify desired reports	.25
6.	terminate computer link	.25
		10-20 minutes

Our scientists are very enthusiastic about the new interactive graphics substructure search system. The number of substructure searches conducted increased significantly after its introduction, and continues to increase. We currently process an average of two substructure search requests a week.

Other Uses

We can conduct combination chemical and biological searches

with our existing search systems. The substructure search system produces a file of the RH numbers of the compounds which met the search criteria. This list is then used to restrict the range of the biological search.

We have also developed search and retrieval programs which allow us to extract connection tables and biological data for use in various data analysis systems.

Summary

We have developed versatile and convenient systems for searching the vast amounts of chemical and biological data generated by our screening programs; significantly increasing the value of the information obtained by greatly facilitating access to it.

Acknowledgements

We wish to thank the Chemical Abstracts Service for providing us with programs and documentation for their Registry II and Substructure Search systems. We appreciate the special attention Mr. Charles Rolle and Dr. Dale Myers gave to us. We also appreciate the guidance offered by Mr. Harold Bewicke, Mr. Richard Dudeck, Mr. Robert B. Smith, Mr. Dugald Brooks and Dr. John Tinker of Eastman Kodak Company on the use of the Registry II system. Finally, we would like to acknowledge the efforts of a large number of people throughout Rohm and Haas who helped ACCIS become a reality: Dr. H.O. Bayer, Mr. Marvin Bell, Mr. Howard R. Boehringer, Mrs. Sandra L. Buckley, Mr. Stephen G. Cohen, Dr. Joseph T. Gilbert, Mrs. Nancy A. Kippenhan, Dr. George A. Miller, Mr. Thomas N. Manuel, Mr. Thomas W. Raezer, Mr. Vernon L. Richens, Mrs. Eileen B. Rothrock, Dr. Alden Spell, Mrs. Marie M. Steinitz, and Dr. Warren H. Watanabe.

Literature Cited

1. Buckley, S.L., Cohen, S.G., Dyott, T.M., Garton, C.R., Gilbert, J.T., Kippenhan, N.A., McNulty, P.J., and Raezer, T.W., paper presented at the 173rd National Meeting of the ACS at New Orleans, March 1977.

2. Spann, M.L. and Willis, D.D., *J. Chem. Doc.*, 11, 43 (1971)

RECEIVED August 29, 1978.

An Integrated Chemical and Biological Data Retrieval System for Drug Development

J. A. PAGE', R. THIESEN, and F. KUHL

Walter Reed Army Institute of Research, Bethesda, MD 20014

The Division of Experimental Therapeutics, The Walter Reed Army Institute of Research, in conjunction with the Division of Biometrics has been engaged in the development and implementation of a large scale integrated chemical - biological data retrieval system for the support of the Army Medical Research and Development Command's drug development activities.

The system is being developed on a Control Data Corporation 3500 with one million bytes of memory, 16 disk drives with removable packs which contain 37 million bytes of storage each, six 7-track tape drives, two line printers, and 16 communication lines supporting line speeds of 110, 300, and 1200 baud.

This effort represents a total redesign of the original system which was described earlier [1].

File Organization

The WRAIR Chemical Information Retrieval System (CIRS) is comprised of four subsystems: Biology, Inventory, Chemistry and the Report Generator. The first three subsystems contain files of information peculiar to each system, and programs for searching these files. The Report Generator is used to combine and sort output from searches of the other subsystems. The subsystems must be searched separately because they are too large to be searched together.

The output from any subsystem may be used to control the search of the next, by means of a common key. For example, a chemistry search yields a number of structures, each of which is identified by a unique accession number. These numbers might then be used to garner information from Biology and Inventory relative to samples of the structures whose numbers come from Chemistry. Or, the list of sample numbers from an Inventory search might be used to extract information from Chemistry

' Present address: Uniformed Services University of the Health Sciences, Bethesda, MD 20014.

relating to those samples. In both examples, the Report Generator would combine the information from the various subsystems
and sort the report into the desired order. The standard CIRS
report may contain information from any of the subsystems alone,
or from any two, or from all three.

Chemistry Retrieval Subsystem. The chemistry retrieval
subsystem file design and organization is being described in
detail for publication elsewhere [2]. Briefly the system consists of two numeric index cross reference files, a screen index
file, and a master structure file. It contains about 270,000
unique structures and occupies 8 disk packs of 37 million characters each.

The three index files provide a cross indexing scheme that
allows for flexibility in sequencing and updating. The system
may be accessed by 1) accession number, a unique number similar
in concept to the CAS registry number, which is assigned by the
chemistry update system to each new structural formula, 2) sample
number, which is a unique, sequential number assigned to each
physical sample without regard to the chemical structure by the
inventory update system, or 3) chemical structures, either whole
or sub-structures.

The accession index file contains the accession number for a
given structural formula and a table of sample index records for
each accession number. This sequence provides quick access to
data for all samples of a particular chemical.

In order to provide continuity and allow for the expression
of a hierarchical relationship the accession number is structured
so that functionally different files may be maintained and salts
may be tied, through their accession number, to the parent
compound. The parts and functions of the parts are as follows:

1)   A two digit alpha prefix designates series. Currently
only two series are being used: "WR" for structures for
which a physical sample has been received for screening and
"XR" for structures proposed or under consideration but not
actually received. An additional series for related structures reported in the literature is planned. The series
prefix is automatically up-graded to "WR" if the compound is
received and processed through the inventory system.
2)   A six digit sequential number which identifies the primary chemical structure.
3)   A two digit numeric "salt suffix" which is assigned by
the update system to different salts of compounds having the
same primary structure. This allows the user to retrieve a
specific compound and all of its salt forms without doing a
sub-structure search. It also allows data on a given compound and all of its salt forms to be grouped together on an
accession number sequenced report.

The sample index file is keyed by the sample or bottle number. This number is cross-referenced to the accession number so that a given sample may be attached to a specific structure. The sequence permits the chemistry subsystem direct access to either the biology or the inventory subsystem and provides for direct access of the structures for reports by the inventory and/or biology subsystems. The sample index file also contains some administrative information about each sample such as the source; the method by which the sample was obtained (e.g. gift, purchased, etc.), whether this sample is the original submission or a duplicate, discreet (i.e., proprietary) or open.

The screen index file is the first file accessed for any structure or sub-structure search. It contains all the information necessary to determine structure matches. When the structure matches have been located, additional information for each structure, such as the structure picture, may be retrieved via the accession number. Because its organization is index-sequential, the screen index file may be accessed either sequentially, or selectively by use of the its indexes.

Each structure has its own record on the screen index. The chief items stored for each are the connection table (in a compressed, non-redundant format), and the structure's unique accession number. The key for each record consists of the accession number, preceded by the structure screen and the partitioning factor. The screen is a 96-bit superimposed code derived algorithmically from the structure. It has been described in detail by Feldman [3]. If two structures have different screens, they must have different structures. Thus only those file structures having the same screen as a given query compound are candidates for matching. Iterative matching will be necessary to confirm the matches, but the amount of iterative matching required is drastically reduced by the screen.

For sub-structure searches, the inclusive property of the screen becomes significant. In the example below, the first structure (discounting hydrogens which are not considered in the calculation of the screen) is wholly contained by the second and as a sub-structure would be a match--

The screen for the first structure is also wholly contained in the screen for the second, i.e. for each bit set in the first screen, the corresponding bit is set in the second. To match a sub-structure then, a candidate's screen must have at least all the bits that are set in the query's screen. The effectiveness of this system in eliminating file compounds from consideration depends on the nature of the sub-structure and varies greatly.

The partitioning factor is conceptually similar to Hode's bucket index [4]. It is a 12 bit code derived from the screen through a series of AND and OR operations in such a way that the inclusion properties of the screen are preserved. Its function in the screen index file is to allot records to one of 4096 partitions with a theoretically uniform distribution. Therefore, in a file of 250,000 records the expected number of records in a given partition is 61. Since for identity searches, the factor and screen of the query must be matched exactly only this very small portion of the file need be read prior to the search. The utilization of the partitioning factor in the sub-structure search is more complex. Any compound containing a given sub-structure will have a partitioning factor which contains at least those bits set in the sub-structure's partitioning factor. As sub-structure queries become more specific more screen bits are usually set, and more bits are set in the partitioning factor. The number of possible inclusive matches to the factor drops exponentially as the number of one bit rises. Because the screen index has the partitioning factor as its major key, it is necessary to read only those records having the right factor. If, however, the sub-structure is so general that more than one third of the partitions must be accessed randomly it is quicker to scan the screen index file sequentially.

The master file contains the chemical structure, which has been captured at input and saved in a condensed form. It is sequenced by accession number as this number is automatically assigned to a new structure by the file update system. In addition to the structure, the molecular formula and qualifiers are in this file. The molecular formula has been stored in such a way as to permit searching both as an exact match or an inclusive match. This format also permits the sorting of matches by molecular formula into CAS sequence for reporting.

Because the connection tables are essentially two-dimensional, and do not contain special bond types, many chemical properties such as stereochemistry cannot be represented. The solution to this and similar problems was the inclusion of machine-readable qualifier fields for each structure to indicate such things as stereo information, polymers, mixtures, and coordination complexes. Each qualifier has also non-searchable free text stored with it to aid human interpretation of the picture.

Biology Retrieval Subsystem. The biology retrieval subsystem consists of two indexed sequential files containing biological test data relating to the structures of the chemistry subsystem. There are over three million records occupying 6 disk packs of 37 million characters each. Both files are sequenced by sample number (BN) and laboratory identification number (Lab I.D.). The data fields are dependent on the type of experimentation done by a specified laboratory and are predefined in a data name dictionary. From a user's point of view the two files are identical. The division is arbitrary and is designed to allow searching only part of the data base. The primary file contains those lab ID's most frequently accessed while the other file contains a number of secondary test systems and historical data. New laboratories can be added to the system by simply adding entries into the data name dictionary under a new lab ID number.

Inventory Retrieval Subsystem. The inventory retrieval subsystem is an indexed sequential file containing information pertinent to the physical samples. It currently contains 433 thousand records and occupies 5 disk packs of 37 million characters each. The file is maintained in sample number sequence. When a sample is received it is assigned the next available sample number and all available data (i.e., date of receipt, source, amount, condition of receipt, shelf location, chemical and physical properties, etc.) are entered into the record. All transactions involving that sample (shipments to testing laboratories, removal from inventory, etc.) and the date of the transactions are also entered into record. The data fields for this file are also predefined in a data name dictionary for searching.

Retrieval Criteria

Chemical Subsystem. The heart of the chemical retrieval subsystem is the sub-structure search capability. The general purpose of sub-structure searching is to retrieve compounds having specified structural similarities. In our system, the similarities are specified in the form of an incomplete structure, which must be included in any file structure that is to be retrieved. While the file structure may contain atoms and interconnections not shown in the query, those in the query must be matched.

Thus, a query sub-structure may contain normal structure atoms and bonds, and indefinite atoms or bonds. The former must be matched exactly, and the latter may be substituted according to the rules governing the particular atom or bond.

Structures, either queries or file compounds, are represented by a connection table. The table contains an entry for each non-hydrogen atom, together with information on the numbers and sizes of covalent bonds on each atom, and the other non-hydrogen atoms attached to it (called "neighbors"). Each entry also shows the number of hydrogens attached to the atom, any ionic charges, and a flag that is set if the atom is in a ring.

We have deliberately discarded the knowledge of what type of bond attaches which neighbor, not because it is uninteresting but because it allows resonating structures, such as phenyl rings, to appear identical regardless of the precise arrangement of double and single bonds. We also make certain adjustments to tautomers to allow them to be identified by either form regardless of which

form is used for input. The details of this have been discussed
elsewhere [5].

Our connection tables are not capable of distinguishing
stereoisomers or polymers. But codes for the presence of such
conditions and text explaining them are stored with the original
input formula and are retrieved with it. These are the qualifiers mentioned above. They may also be specified for a chemistry search.

Normal structures are coded for input by means of a special-
ly-modified teletype [6], which allows the structural formula to
be typed as a combination of atoms and bonds to represent chains
and rings. It also allows strings of subscripted element symbols
and groups inclosed within parentheses, whose connections must be
inferred. The extensive logic necessary to interpret these formulas has been described [7]. The result is a fairly simple set of
rules for the chemist writing the formula for input that generally corresponds to normal chemical conventions. These structures
are complete, and except for certain Markush-type structures, no
ambiguity or substitution is allowed at any point. Queries, on
the other hand, are incomplete structures, allowing additions and
substitutions at specified points.

Most query atoms are normal atoms. That is, they are not
special atoms, they have no unspecified bonds and their valance
is not zero. It is required that they match file atoms. The
file structure must contain one identical atom for each normal
atom in the query. If the query atom is in a ring a flag is set
in the atom descriptor word and the file atom must also be in a
ring. However, if the query atom is not in a ring, the file atom
need not be in a ring, but it is allowed to be. For example:

$Z-NH-Z$ will allow $HN\langle\hexagon\rangle$ and $CH_3-NH-CH_2CH_3$

If you wish to force a query atom to be matched only by a file
atom which is a ring member, the query atom must be in a ring.
If the query cannot be written in such a way as to include the
particular query atom in a ring, ring members may perhaps be
specified with a special atom.

The simplest type of substitution is that of a special atom.
They appear in the query as atoms with special symbols and may be
replaced in the file struture by any atom meeting the criteria
they impose. In Figure 1, the query structure contains two special atoms, an "X" which allows the substitution of any non-hydrogen atom and a "Q" which allows the substitution of any non-
hydrogen, non-carbon atom. These special atoms may appear anywhere within the query structure, that is, they need not be terminal atoms but may be incorporated in a string or in a ring. In
order to be a "hit" to a query a file atom need only have at
least the bonds, charges etc., indicated for a special atom.

Therefore, the file compounds in Figure 1 contain many atoms
other than the normal atoms and the special atoms in the
query. The special atoms allowed and their restriction on ring
membership are given in Table 1.

Table I.
Types of Special Atoms

Symbol	Atom Type	Ring Membership
Z	Any	Indifferent
X	Not H	Indifferent
Xr	Not H	Required
Xc	Not H	Excluded
Q	Not H, not C	Indifferent
Qr	Not H, not C	Required
Qc	Not H, not C	Excluded
Ha	F, Cl, Br, I	Indifferent
Mt	Any metal	Indifferent
Rc	Carbon	Required
Cc	Carbon	Excluded

Any atom in the query, including special atoms may carry a
charge, which must then be matched by the file atom. The inverse
however, is not true. That is, the absence of a charge on a
query atom does not preclude a charged file atom as a match.
Because the valence of special atoms is indeterminate (except for
Ha, Rc, Cc), all specifically required bonds to a special atom
must be shown explicitly. The ring and chain carbons (Rc and Cc)
may be used to allow one to write part of a ring as a chain, or
to exclude rings, especially fused rings, as answers. (See
Figure 2).

Figures 3,4, and 5 show the relationships among the special
atoms and how they may be used to modify a query so that it
becomes more general or more specific depending on the nature and
the number of matches desired. Figure 3 divides the universe of
possible substitutions into eight categories and lists which of
these categories will be retrieved by each special atom. Figure
4 shows representative structures for each category singly substituted on a methyl group and which of these structures would be
retrieved by each special atom. Figure 5 shows a few of the
possibilities for an ortho di-substituted phenyl ring. These
simple examples give, of course, only an indication of the versi-
tility possible. Combinations of these few special atoms provides a very powerful sub-structure search capacity.

The other major query element is the unspecified bond, written as any bond overstruck with a question mark. In general, the
unspecified bond may be used to allow the connections on the
connected atoms to vary, as long as the neighbor relation is
maintained. Used between two normal atoms, it requires the two

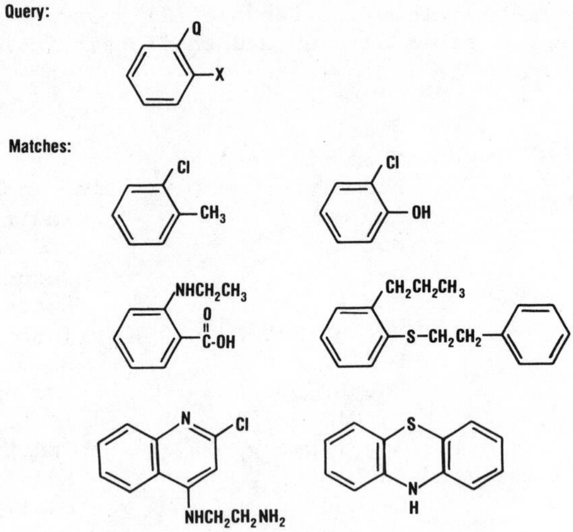

Query:

Matches:

Figure 1.   Example of a simple substructure query

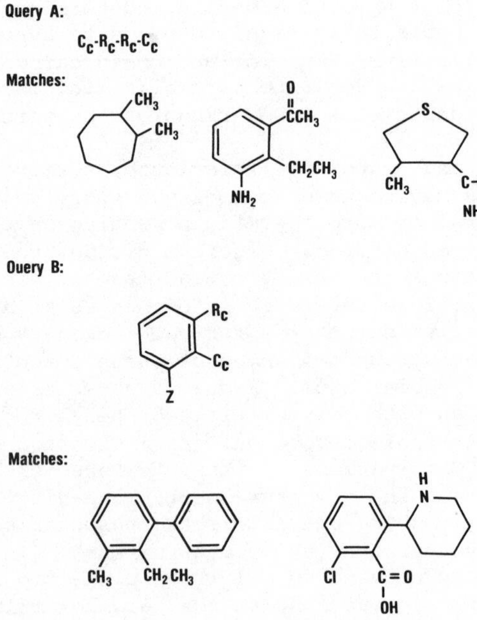

Query A:

$C_c\text{-}R_c\text{-}R_c\text{-}C_c$

Matches:

Query B:

Matches:

Figure 2.   Substructure query using R_c and C_c

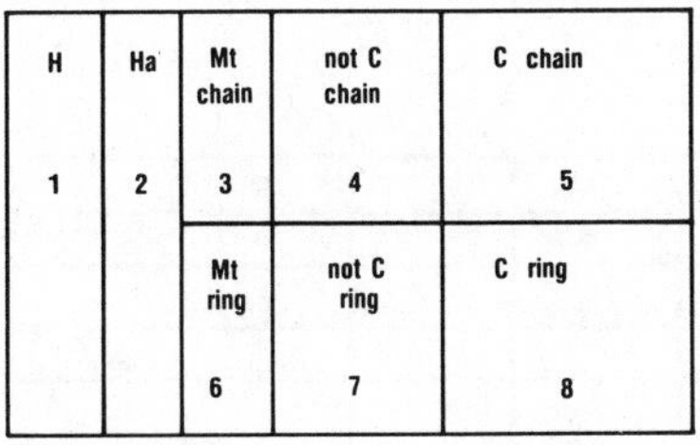

H	Ha	Mt chain	not C chain	C chain
1	2	3	4	5
		Mt ring	not C ring	C ring
		6	7	8

Z = 1-8   Qr = 6-7
X = 2-8   Ha = 2
Xc = 2-5   Mt = 3,6
Xr = 6-8   Cc = 5
Q = 2,3,4,6,7   Rc = 8
Qc = 2-4

Figure 3. Areas of substitution allowed by the special atoms

CH_3-

Query → Response ↓	Z	X	Xc	Xr	Q	Qc	Qr	Ha	Mt	Cc	Rc
CH_4	•										
CH_3-CH_3	•	•	•							•	
CH_3-⬡ 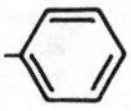	•	•		•							•
CH_3-Cl	•	•	•		•	•		•			
$CH_3-Cd-CH_3$	•	•	•		•	•			•		
CH_3-NH_2	•	•	•		•	•					
CH_3-N⬡ 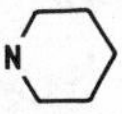	•	•		•	•		•				
CH_3-Sn 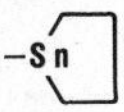	•	•		•	•		•		•		

Figure 4. Substructure substitution at a single point

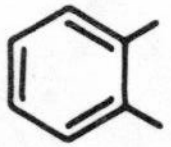

Query → / Response ↓	Z: Z	Z: Q	X: Z	X: Q	Xc: Z	Xc: Q	Xr: Z	Xr: Q	Q: Z	Q: Q	Qc: Z	Qc: Q	Qr: Z	Qr: Q	Ha: Z	Ha: Q	Mt: Z	Mt: Q	Cc: Z	Cc: Q	Cr: Z	Cr: Q
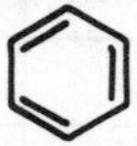	●																					
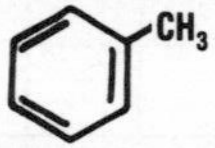	●		●		●														●			
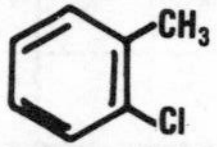	●	●	●	●	●	●			●		●				●				●	●		
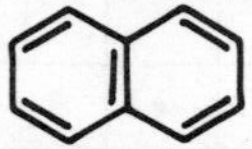	●		●				●														●	
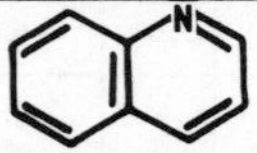	●	●	●	●			●	●	●				●								●	●
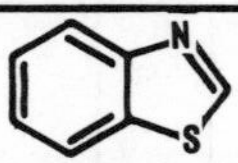	●	●	●	●			●	●	●	●			●	●								
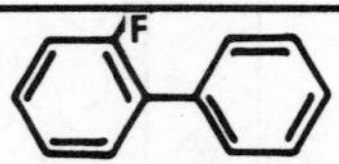	●	●	●	●	●		●	●	●		●				●						●	●
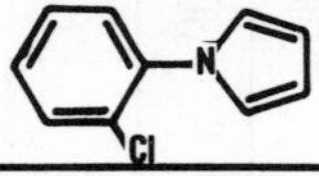	●	●	●	●	●	●	●	●	●	●	●	●	●	●								

Figure 5. Substructure substitution at two points

atoms to be neighbors, but allows the valence of the two atoms to vary (Figure 6) or the nature of the attachment between the atoms to vary (Figure 7).

A single query may consist of a sub-structure fragment with normal atoms and any number and combination of special atoms and unspecified bonds. It may in fact consist of only special atoms. It may also consist of several sub-structure fragments each of which may have special atoms and unspecified bonds. These fragments may be related to each other in any combination of three ways. If the fragments are simply disconnected, each fragment must be distinctly and simultaneously present in the file structure. If however two fragments are related through the Boolean operator "AND", they are matched independently. Thus if parts of the fragments are identical, those identical parts are redundant and need not be distinctly present. For example:

$$X - \langle \text{ring} \rangle - X \cdot X - CH_2 - Cl \cdot Cl$$

requires at least two Cl atoms in the response.

$$Cl - \langle \text{ring} \rangle - CH_2Cl \qquad NO_2 - \langle \text{ring} \rangle - CH_2CHCH_2Cl$$
$$\overset{|}{Cl}$$

while

$$X - \langle \text{ring} \rangle - X \text{ AND } X - CH_2 - Cl \text{ AND } Cl$$

allows two Cl atoms in the response but only requires one.

$$HO - \langle \text{ring} \rangle - CH_2Cl \qquad ClCH_2CH_2 - \langle \text{ring} \rangle - NHCH_2CH$$
$$ClCH_2CH_2 - \langle \text{ring} \rangle - \overset{O}{\overset{\|}{C}}Cl$$

etc. in addition to the first responses.

Two fragments may also be related through the Boolean operator "AND NOT". In this case the file structure must <u>not</u> match the specified fragment. Each query may contain up to 32 such fragments in any combination that is not directly contradictory.

<u>Biology and Inventory Retrieval Subsystems.</u> Essentially the same programs are used to search the biology and inventory systems. The major difference is in the data name dictionary that is attached. Each subsystem has a dictionary of all data items in its file. This provides the capability of searching on any field or combination of fields in either data base. A field may be defined as numeric, alpha, alphanumeric or as a repeating group. Also additional flexibility is provided by allowing new fields that are not in the data name dictionary to be defined and used in the search. The search is made up of a search command (SUBSET, or SUBSET, USING), the field(s) to be selected as defined

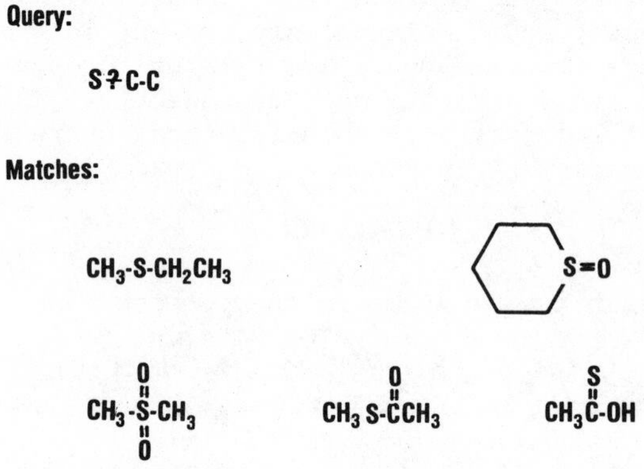

Query:

$$S \overset{?}{-} C\text{-}C$$

Matches:

$CH_3\text{-}S\text{-}CH_2CH_3$

$S=O$

$CH_3\overset{O}{\underset{O}{\overset{\|}{\underset{\|}{S}}}}CH_3$ $CH_3 S\overset{O}{\overset{\|}{C}}CH_3$ $CH_3\overset{S}{\overset{\|}{C}}\text{-}OH$

Figure 6.   Unspecified bond query

Query:

$$N \overset{?}{-} N$$

Matches:

$N{\equiv}N$ $CH_3CH{=}N\text{-}NHCH_2$

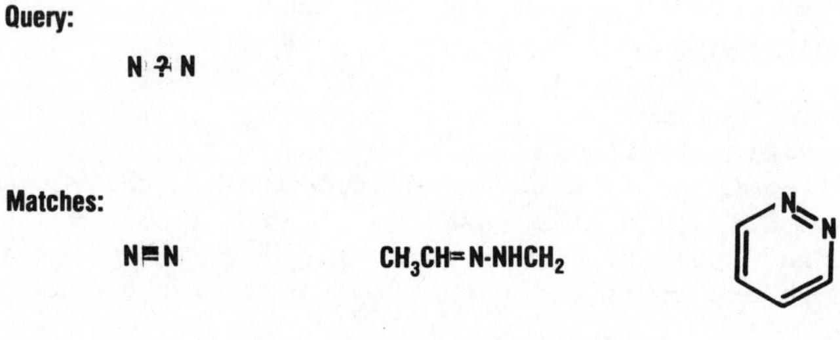

Figure 7.   Unspecified bond query

in the dictionary, and the search criteria. The retrieval system
is capable of handling up to ten imbedded ANDed or ORed functions
as search criteria. There are five operations which are used with
the data fields to make up the search criteria. They are EQUAL,
GREATER THAN, LESS THAN, CONTAINS (which compares for a specific
string of characters) and the NOT of each. The CONTAINS operator
used with the word KEY works as a partition search using only the
major part of the key. Thus it functions as a generic search.
Searches using repeating groups must use a subscript in the data
items declared as repeating in the dictionary. For a specific
occurance of a repeating group, a number subscript is used in the
search criteria. The subscript ALL is used when all occurrences
must meet the search criteria.

Searching Procedures

Chemistry Subsystem. The chemistry system may be searched
in either an on-line or batch mode. Each mode has differing
capabilities and uses.
The on-line system will perform 4 distinct types of searches: 1) by accession number; 2) by sample number; 3) by full
structure, or identity search; 4) simple sub-structure search.
All on-line functions use an IMLAC PDS-4 intelligent graphics
terminal. The IMLAC machine contains a small computer with 8K of
memory, and a display processor driving a graphics CRT. Structures are represented in the IMLAC by the character set developed
for the chemical teletype [6]. This character set allows two-
dimensional representation of most structures. Each character,
along with the blank, the backspace, the line feed and the reverse line feed, has its own 7-bit ASCII code, and a picture is
stored in the IMLAC as a series of such codes. The IMLAC processor is programmed to interpret the codes and draw the corresponding characters on the CRT.
The IMLAC is also programmed to allow its operator to enter
or edit such pictures with its keyboard. Pictures may also be
sent or received to or from the host computer. Thus, answers
may be displayed to the IMLAC operator by transmitting the ASCII
representation of the structure via phone line to the IMLAC. The
IMLAC could in fact be used for primary input rather than the
teletype and in a system with less thruput this would be desirable. The teletypes are used because they are much less expensive, we already owned them, and the operators are familiar with
then.
The first three types of on-line retrievals listed above
were designed to replace manually-searched card files. With the
program, the operator may retrieve and display structures by
knowing either the accession number, the sample number, or the
structure itself. In the case of an identity structure search,

the index-sequential design of our connection-table file allows
the structure to be retrieved in less than 5 seconds, and the
picture to be displaced in another 5 seconds. Since the structure screen is computed from the structure itself, only the
structure must be entered for identity queries. The person
entering the query need not know the index criteria.

The on-line sub-structure search was intended originally as
a tool for testing the system, but has been made available to
selected users of the system. Only the simplest searches are
allowed: There is no Bollean combination, nor is there any ability to search any information except the structures. (This
means, among other things, that the entire data base is eligible
for the report, regardless of the privacy attaching to any compound. Since discreet compounds are thus available, access to
the program must be restricted.) The full range of special atoms,
unspecified bonds, and structure qualifiers is available to the
on-line user. Answers are returned by structure and accession
number only (not by sample number), each structure being displayed as the user asks for it. If the corresponding sample numbers
are desired they can be obtained by requesting all sample numbers
for the given accession number.

The usefulness to the WRAIR Staff of searching for sub-
structures on-line has yet to be determined. Since the chemistry
retrieval system uses the screen and connection table generated
by the query to search to screen index file for either a whole
or a sub-structure no more information (disk-drives) is needed
on-line for sub-structure searching than for identity matches, so
one is no more costly than the other, in terms of disk resources.
But the elapsed time needed for such sub-structure searches
varies greatly and may be large. The time needed may be kept
small if the query is specific and many screen bits are set.
Then, because the files are indexed by an abbreviation of the
screen, i.e., the partitioning factor, they need not be read
completely, thus eliminating most of the input/output time necessary to read the entire file. (Of course additional time will
also be saved because such a situation will limit the number of
iterative matches needed as well.) The worst situation is a
query so general that the entire screen file must be read, and
most of it must be searched iteratively. The time for usch a
case depends largely on the size of the file and the competition
for CPU time from other multi-programmed jobs. The estimated
worst-case search would require about 35 minutes elapsed time in
an otherwise empty machine, for a file of 250,000 structures.

The economies to be gained by searching many sub-structures
at once, i.e. running batch searches, are great. They arise
chiefly because the time needed to read the file of structures
from the disk may be apportioned among the sub-structure queries.
This time is larger than the time required by the iterative
searching usually needed for searches, and thus the time needed
to search for a batch of 10 sub-structures is no where near the

time needed to do 10 single sub-structure searches. For example
10 batched general queries of the worst-case typed noted above
should require only about 45 minutes of elasped time.

There are two additional reasons for a batch off-line
search. First, the search criteria can be expanded to allow
searches that are not possible on-line because of time considerations, as in the case of Boolean combinations of simple
fragments, or available computer resources, as in the case of
criteria based on non-structural information that is stored on
other files. The second reason for batch searching is based on
the need to efficiently handle large numbers of queries and
answers which require a source of high-volume hard copy output.

All searches, whether for on-line or batch processing, may
be formulated interactively. The structure fragments are generated on the IMLAC terminal and edited on-line. If erroneous,
they are immediately returned for correction. Thus the user may
define and edit all the questions for a given batch search on-
line, and be sure they are at least in the correct syntax. This
capacity increases the effectiveness of the batch search by eliminating long delays in turn-around due to input errors. Individual fragments, with no Boolean combinations, may be used for a
preliminary search on-line. Thus the user has the capability of
formulating a query and looking at a predetermined number of
answers. Based on this "preview" the user may then submit the
query to the batch search, redefine the search criteria and look
at a revised set of answers, or delete the query entirely. This
preview capability may in fact prove to be the most useful
function of the on-line sub-structure search.

The batch search also contains the capacity to search by a
specified molecular formula. This capability was not included in
the on-line search because there was no requirement for it.

Batch search output is printed off-line on a Versatec electrostatic dot matrix printer/plotter. It prints 100 dots per
inch and uses 11 inch wide paper which is easily incorporated
into reports.

Biology and Inventory Subsystems. These subsystems are only
searched in batch mode. Quick access is provided by periodic COM
listings in accession number and sample number sequences. Because of the size of the files, simple queries (i.e. what is the
malaria screening data for compound A or how much of compound B
do we have on hand) are most efficiently and economically handled
with microfilm. More complex searches and searches requiring
interaction among the sub-systems require sufficient computer
resources to prohibit running during prime time on a multi-
programmed computer.

While either the Biology or Inventory subsystems may be
searched independently, in practice this is seldom done. More
often, information from more than one system must be found in

order to answer a query. If the search begins with the Biology
system the SUBSET command provides the interface with the other
systems by generating a file of answers with can be used as
input to continue the search. For example a search such as

     SUBSET, BIOREC, WHERE, BIOSOURCE = 2300
     AND, BIOPGS = G, OR, BIOSOURCE = 2300,
     AND, BIOPGS = P.

would create a file of biology records where the source of
the compounds was 2300 and the type was either P (purchased) or G
(gift). This file could then be used, for example, to find
additional samples of the same compounds from other sources. The
SUBSET, USING command permits the output of one search to be used
as the controlling factor to a second search. A record is read
from the using file, the key is extracted from the record and is
used to perform a random find on the new input file. Once the
record is found the additional criteria, if any, are applied and,
if these criteria are met, a new subset file is created from the
matched records. For example:

     SUBSET, USING, FILA, CIS, WHERE,
     AMTON= 500 M, AND, ARRAY (ALL) NOT = MM.

The previously retrieved biological data sub-file would be
used to query the inventory file. Once matching records (i.e.
records with the same sample number) were found they would be
checked for an amount on hand equal to or greater than 500 mg and
no shipment to test system MM. If these criteria are met, the
matching inventory records are used to create a new SUBSET file.
The search may then continue to obtain additional information
from the chemistry files. The Report Generator would then process all appropriate SUBSET files according to criteria specified by the user to generate the final report.

Search Strategies

     As indicated earlier, most searches require accessing more
than one data base. There are no rules governing the sequence in
which the different subsystems must be searched. The output
from a search of any system can be used, through the SUBSET USING
command, as criteria or partial criteria for searching any other
subsystem. The outcome of any given search should be the same
regardless of which system is used to begin the search. However,
the amount of computer time required to complete a given search
may in fact depend on the sequence in which the systems are
searched. In general it is best to begin a search on the system
which will be most restrictive in the number of responses.
However, the determination of that system is not always obvious
and requires a fair degree of familiarity with the contents as
well as the organization of the data bases. For example, suppose
a user wishes to send to test system B all quinoline derivatives
that are active in test system A. In addition, test system B

requires a minimum quantity of 700 mg to complete the test and
the user does not wish to deplete the total inventory supply.
The search criteria then are:
    Biology subsystem search:
        SUBSET,BIOREC,WHERE,LABID=TSA.
    Inventory subsystem search:
        SUBSET,INVREC,WHERE,LABID(ALL),NOT=TSB,
        AND, QUANT, NOT   1000 mg
    Chemistry subsystem search:
    Structure=

   The user must realize that starting this search with the
inventory system will not provide a satisfactory answer since the
inventory system is sequenced by sample number. Even though an
individual sample may appear to meet the inventory criteria it is
necessary to know the pertinent accession numbers so that inventory data for all samples of a given compound may be compared
against the inventory criteria together. That is, if any one
sample of a compound with multiple inventory entries has been
shipped to TS B then all samples of the same compound, regardless
of shipping data, should be rejected by the search. The search
cannot do this until it has the necessary information to correlate sample data. Similarly the decision to begin with either
of the remaining systems rests on the user's knowledge of the
data. That is, if there are a large number of quinolines on the
structure file and test system A is a small laboratory with
relatively few actives, it would be preferable to start the
search with the biology system. If the sub-structure criteria of
the chemistry portion of the search are relatively specific then,
because of the speed of the search, it may be desirable to start
with the chemistry search.
   The feasibility of programming a master search which would
accept free text search criteria from the user and construct a
search strategy is being studied. First however, more user experience in the development of effective search strategies is
necessary.
   The Report Generator contains additional options open to the
user which are not precisely search criteria. Because of the
proprietary nature of much of the data base, unless the user
specifically indicates otherwise, only open (non-proprietary)
data will be reported. It is also possible for the user to
specify all hits on a combined search be reported rather than
only matched hits. For example the chemistry data base could be
searched for triazines and the output used to search the biology
data base for malaria actives. The user has the option of specifying that only those triazines that show malaria activity be

reported or that all triazines along with any available data
showing malaria activity be reported.

The user may specify that only data within a certain range
of accession numbers or sample numbers be reported, or that only
data received within a specific time span be reported. In addition the user may specify the sequence of the data in the
report.

Applications

The CIRS is used by the Division of Experimental Therapeutics in support of a large drug development program. The
Retrieval System (as separate from the structure registery
and data maintenence system) is routinely used for a variety
of functions.

Analysis of Research Proposals. The Army Medical Research and
Development Command supports research in the directed synthesis
of screening candidates. Twice yearly the Division of Experimental Therapeutics reviews synthesis proposals. The proposed
structures in each proposal are entered into the data base in the
"XR" series and are also entered as identity searches to determine whether or not the compounds are already on hand and to
identify duplication among the proposals. In addition, substructure searches for the major classes of compounds proposed
are also run. Matches from both the identity and sub-structure
searches are then used to query the biology and inventory files.
The reviewer is presented with a report sequenced by proposal
number providing him with all available information on the availability and activity of all specific compounds and classes of
compounds in each proposal. A similar procedure is used by contract monitors to review progress on synthesis contracts and to
prevent duplication of effort.

Review of Screening Data. Structures are added to the screening
data after the data are received from the laboratories. The
screening data simply act as whole structure queries to the
chemistry retrieval system. The report is sequenced by accession
number so that data for duplicate samples will be grouped together. Sub-structure searches can be formulated from the report
to identify additional samples either on hand or in preparation
in those classes of compounds that are interesting. The matches
from the sub-structure search can then be used to further query
the biology and inventory files to determine if those additional
samples identified have already been screened, if so their activity, if not the amount available, and a possible source for
obtaining additional material.

Monitoring of Screening Laboratories. The system is used to
determine correlation of activity between primary and secondary

screens for the same disease and between basic screens for different diseases. It is also used to determine samples with sufficient quantity for screening when a new test system is being developed. It can be used to moniter ongoing activity. For example the user could request a report on all samples shipped to a given test system for which no biological data has been reveived.

Conclusion

The chemical-biological data system has been an integral part of the Army's drug development program for over 12 years. The current up-grading of the original system will eliminate the labor-intensive maintenance of manual files, allow closer monitoring of all aspects of the program and provide information in a more timely manner. Those features of the system which are particularly attractive to the end user are structure input and output, machine coding of sub-structure queries, on-line editing of queries, and user designed reports which provide correlated data from all data files.

References

1. D. P. Jacobus et. al., "Experience with Mechanized Chemical and Biological Information Retrieval Systems." J. Chem Doc Vol 10, p 135, 1970.

2. J. A. Page, R. Theisen, F. Kuhl, Manuscript in preparation.

3. A. Feldman, "An Efficient Design for Chem Structure Searching. I. The Screens." J. Chem. Info. and Comp. Sciences, Vol 15, No. 3, 1975.

4. L. Hodes and A. Feldman, "An Effective Design for Chemical Structure Searching. II. File Organization". J. Chem. Info. and Comp. Sciences, in press May 78.

5. A. Feldman, "An Efficient Design for Chemical Structure Searching. III. The Coding of Resonating and Tautomeric Forms. J. Chem Info and Comp. Science, Vol 17, No. 4, 1977.

6. A. Feldman, "A Chemical Teletype." J. Chem. Doc., Vol 13, No. No. 2, 1973.

7. A. N. De Mott, "Interpretation of Organic Chemical Formulas by Computer", IEEE Spring Joint Computer Conf, 1968, p 61.

RECEIVED August 29, 1978.

13

The Drug Research and Development Chemical Information System of NCI's Developmental Therapeutics Program

SIDNEY RICHMAN, GEORGE F. HAZARD, JR., and ALICE K. KALIKOW

Developmental Therapeutics Program, Division of Cancer Treatment,
National Cancer Institute, NIH, Bethesda, MD 20014

The Developmental Therapeutics Program (DTP) in the Division of Cancer Treatment, National Cancer Institute (NCI) is responsible for all preclinical phases in the development of cancer therapeutic agents. As one approach to identifying leads to such drugs, the DTP (which subsumed the Drug Research and Development Program(1)) acquires drugs, synthetic materials and natural products and tests these materials in a variety of antitumor systems. Since the inception of the NCI cancer chemotherapy activity in 1956, some 316,000 chemicals and drugs have been screened in approximately 4,000,000 animal and in-vitro tests. Currently, 15,000 such materials are acquired and evaluated yearly.

Evolution of the System

As in other biological screening programs, the supporting chemical information system (CIS) must provide, at a minimum, two basic capabilities: the registration of acquisitions and access to the data base. The chemical system which was started in 1956 employed a manual registration procedure based on cataloging compounds by molecular formula and differentiating between compounds having the same molecular formula by the systematic chemical name. Systematic names for accessions were maintained on punched paper tape. Substructure searching for analogs was based on chemical fragment codes assigned to accessions and an optical coincidence system used for retrieval.(2) The second system, used from 1967 through 1974 consisted essentially of three components: the Chemical Abstracts Service (CAS) Registry System (3,4,5) as adapted for NCI private use; CAS' automated batch substructure search procedures (6,7); and computer-generated indexes for identifying substances by systematic name.

The passage of the National Cancer Act of 1971 resulted in an increased yearly acquisition rate which rose in three years from a level of `10,000 chemicals and drugs to 39,000. With a file size of 158,000 substances and complex manual procedures required despite the automated processing of accessions described above, it was evident that there was need for a CIS more responsive to operational requirements. In June, 1971 the University of Pennsylvania, under an NCI contract, began the design and implementation of DTP's Drug Research and Development Chemical Information System (DR&D CIS) with additional programming support provided by E.I. Dupont de Nemours and Company under subcontract. In July, 1974, this new system was installed at the Chemical Abstracts Service (CA) which operates the system under an NCI contract. The INVENTORY and SPECIAL STRUCTURAL FEATURE IDENTIFICATION processors were added to the CIS in February, 1976 and September, 1977 respectively.

Design Constraints

The constraints on the design were to avoid research, use existing systems or programs wherever possible, and be compatible with the CAS environment to facilitate operation of the existing and planned CAS chemical information handling techniques. The following components were obtained from other systems and were integrated (sometimes with modifications) into the total system: CAS Data Management System (FIDO)(8);   CAS Registry III (9,10,11); CAS Transaction Editor; DCRT (NIH) On Line Structure Input System for query formulation (12); General processing utilities and a macro processor from Dupont; U.S. Army CIDS Substructure Ring Screens and Interactive Search (13); CAS' Algorithmic Structure Display (ASD) (14); and CAS' On Line Structure Input System (15) for input to Registry.

Interaction of the CIS with Accessioning and Screening

Figure 1 outlines the routine interaction of DTP's chemical and drug accessioning and screening activities (which are heavily contract supported) with the chemical and biological data processing systems. Samples of compounds are submitted by suppliers or are collected at suppliers' sites. Accompanying the samples are data sheets which include the structural and supplier identification and occasionally biological and physical properties. The Acquisition activity weighs and temporarily stores the sample, prepares and transmits the input documents (which include structural and non-structural data) to the Chemical Information Processing activity, and awaits the return of 21 action and informational system outputs. Reports are forwarded to suppliers advising them of the accession numbers assigned to their submissions or requesting the replenishment of supply for a previous submission to complete testing. Bottle labels are

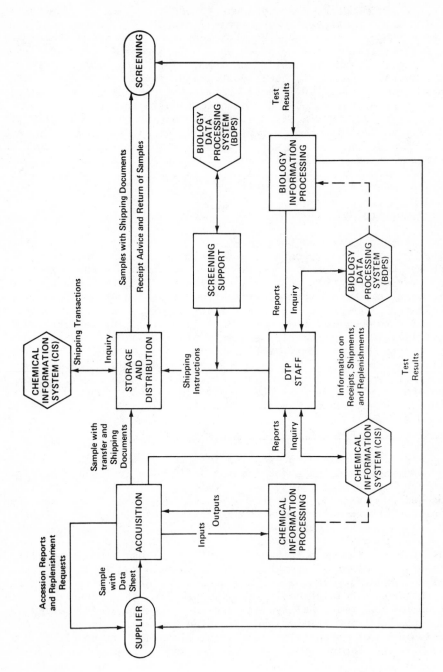

Figure 1. Interaction of accessioning and screening with data processing

affixed to the samples which are transferred to the Storage and Distribution activity with advice on conditions for storage and shipping and with computer-generated manifests for shipping samples to Screening Laboratories for routine biological testing. Among the variety of outputs used by the DTP staff is the request that special testing be scheduled for a sample identified by the system to be an analog of program interest. (See Table 1 for a description of selected system outputs.) The CIS transfers to the Biological Data Processing System (BDPS) information in a machine-processable form regarding the receipt of samples (both new to the collection or duplicates of previous acquisitions), shipments made, and replenishment actions. The Screening Support activity queries the BDPS to secure and review prior testing for newly-received duplicates and refills and to decide on the need for additional testing.

When the Storage and Distribution activity receives manually-prepared shipping manifests for the shipment of analogs of Program interest for special testing or for reshipments, the chemical file is queried to secure the quantity remaining in stock, storage location, and special instructions regarding the conditions of shipment, storage, and testing. Transactions regarding shipments and returns are input and edited on-line; however, the file is updated in a batch mode. Records of action to secure an additional supply of a compound are input and communicated to the BDPS as well as advice on the receipt of refills

The Screening laboratories acknowledge the receipt of shipments, test the compounds, forward the results to the BDPS, and return the excess supply to the Storage and Distribution activity. The BDPS outputs summaries of the test results which are distributed to suppliers, screeners, and the DTP staff.

File Structure

Figure 2 illustrates the relationship among the principal files of the system, as well as the hierarchical structure of the NSC Cross Reference file. The NSC Number is a unique serial number machine-assigned to each compound. Because of confidentiality requirements, a different NSC Number is assigned to a substance submitted by an industrial supplier which has the same structure as a sample previously submitted by another industrial supplier. The structure representing the compound is registered by the NCI version of the CAS Registry III System only once under a separate but unique registry number called the IRN (Internal Registry Number, not to be confused with CAS' Registry Number) and the connection table is stored in the NCI Registry Structure file (and the Registry Ring file). All the other data pertaining to a compound submission, including the IRN, is in the

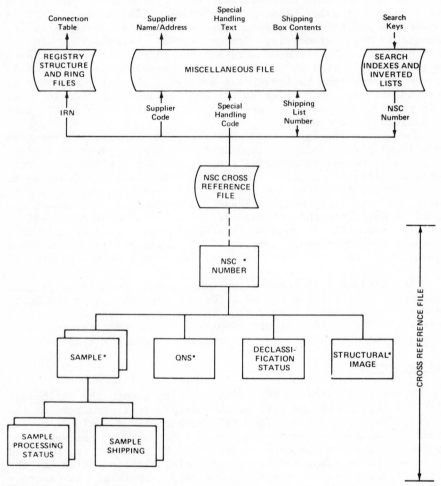

Figure 2.   File relationships in the DR & D CIS

Cross Reference file. The connection table in the Registry
Structure file is accessed for a given NSC Number by the IRN
present in the NSC record of the Cross Reference file. The name
and address and other control information for a given supplier in
the Miscellaneous file is accessed by the supplier code in the NSC
record for a given compound. The NSC Numbers of all the compounds
scheduled for shipment on a specific Material Shipping List (MSL)
is accessed for a specific carton of samples by the MSL number in
the MSL record in the Miscellaneous file. As will be described
later, the INQUIRY subsystem uses the search indexes and inverted
lists to be lead to NSC Numbers and their associated data in
response to full structure or substructure searches. (See Table 2
for the size and growth rate of the DR&D CIS files.)

The hierarchical NSC Cross Reference file has at its highest
level the NSC Number record, containing only NSC-related data such
as IRN, confidentiality, CAS Registry Number, designation of
antitumor activity, compound class, etc. The second level of the
hierarchy contains four record types. The Sample record relates
to specific samples of the NSC, containing for each sample such
data as supplier code, supplier compound identification and lot
number, weight, etc. The QNS (Quantity Not Sufficient) record
relates to replenishment actions when additional material is
required to complete testing. The Declassification record deals
with actions pending regarding the declassification of
confidential submissions. The Structure Image record contains the
molecular formula and the structural image of the compound which
is stored only once for a group of NSC records having the same
structure. The third level of the hierarchy below the Sample
record has two record types, each of which may have multiple
records. The Sample Processing Status record contains information
regarding the status of a given sample as it traverses through the
DR&D CIS. When the processing of the sample is complete, the
status record is deleted. The Sample Shipping record contains
data regarding each shipment of the sample, such as the amount
shipped and to whom.

The System as a Network of Processors

As illustrated in Figure 3, the DR&D CIS may be represented
as nodes in an interconnected network. The blocks in the figure
represent computer jobs (each with one or more processing steps)
that are called processors. A processor receives transactions via
manual entry through an editor (E inputs) or via another processor
(P inputs). The application program (or programs) within a
processor relates to a specific set of functions performed on a
series of transaction types. Associated with each processor are a
variety of man/machine interactions which may be functionally and
organizationally subdivided further. Such subdivisions are called
stations, analogous to work stations in a production plant.
Twenty-one work stations are mapped on to processors in the

Table 1. Selected System Outputs and Their Use

B and 1D Record Journals — data automatically transmitted to update the BDPS with data regarding shipments, testing scheduled, second suppliers, and replenishment actions.

Biology Inquiry Listing — data on duplicate samples and refills used in the BDPS to secure test results from previous samples and to schedule additional testing.

Bottle Labels — used to identify the contents of the shipping and reference bottles.

Chemistry Card — basic manual reference for accessions, including information such as structural diagrams and chemical class of interest.

Drug Transfer Record — manifest of materials to be transferred from Acquisition activity to the Storage and Distribution activity, including instructions for distribution.

Inventory Exception Report — advice of the need to replenish the supply of positive control compounds when reorder levels are reached.

Material Shipping List — manifests of compounds to be shipped to Screeners for routine screening.

Name and Address Card — reference tool containing the name and address and other identifying information of suppliers and screeners.

NSC Number List — advice to suppliers of the NSC (accession) Numbers assigned to their submissions, including structures.

Refill Report — periodic report of refills received to assure prompt shipment.

Request for Manual Material Shipping List — request for scheduling special testing of a material identified to be an analog of Program interest.

Request for Declassification — request for a supplier to authorize the release of information of a confidential sample, e.g., for publication.

Screener Receipt Overdue Report — to assure the receipt of shipment to screeners.

Table 2. DR & D CIS File Sizes[a]

File Name[a]	Current Size (Megabytes)	Annual Increase (Megabytes)
Cross Reference	191.0	17.0
Extended Molecular Formula Index	16.5	0.8
Ring ID	2.9	0.2
Miscellaneous	2.4	0.2
Statistics	0.2	0.0
Chem Search Index	1.9	0.1
IRN/NSC* Index	19.1	0.8
NSC* Inverted List	4.6	0.2
NSC* Bit map	11.6	0.5
Registry Structure	96.7	4.5
Registry Ring	5.9	0.5
Transaction File	100.0	0.0
TOTAL	452.8	24.8

[a] In addition to the listed files which reside on disk, two master files reside on tape; the NSC*/Key file is contained on eight reels and the Key/NSC* is contained on two reels of tape. The increase in file size is based on the accession of 15,000 new compounds annually.

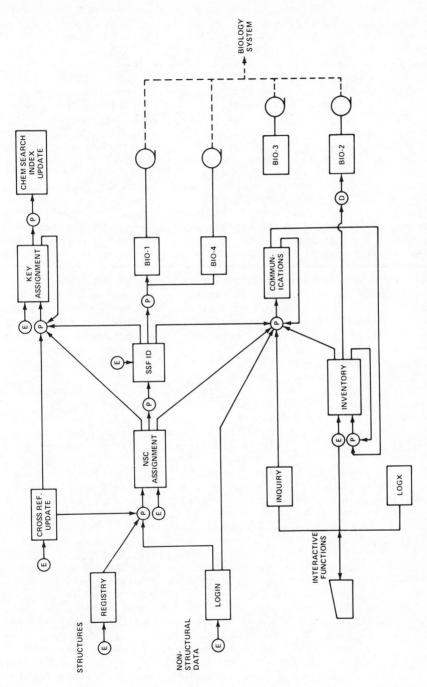

Figure 3. Processor network of the DR & D CIS

system. (For example, the NSC Assignment processor has three such
work stations: Declassification Decision, requiring communication
with suppliers; NSC Decision, requiring DTP action based on the
review of information regarding duplicates; and NSC Assignment, a
programmed decision for most cases based on the uniqueness of the
structure, supplier information, and confidentiality.) Data
regarding samples are processed through different paths depending
on the conditions of the submission (e.g., confidential versus
non-confidential) and characteristics of the compound itself
(e.g., structure new to the file versus a duplicate of a previous
accession). Thus, a new submission must go through Registry but a
refill does not. A particular traversal through the stations of
the system is called a journey. The DR&D CIS has fourteen
processors, twenty-one work stations, and thirteen journeys. A
system control program called TRAC monitors the processing of
sample data through the various stations required by the differing
journeys. For every journey type in the system, TRAC maintains
the Sample Processing Status record in the NSC Cross Reference
file which contains the combination of status codes for each of
the stations appropriate for the journey of the accessed sample
through the system. Every submission is examined periodically by
a program which reads its status record, compares the current date
with its expected completion time at each station and reports any
that is overdue. A second control mechanism lists every sample in
process as an array of status codes at each appropriate station
for each journey. These reports are used to oversee and manage
the system by identifying existing and potential bottlenecks.

A brief description of the basic programs or processors
follows. LOGIN - inputs non-structural information regarding
compound samples. STRUCTURE INPUT AND EDIT - converts to a
connection table structural data input by a chemical typewriter or
other input devices and edits the structural diagram for chemical
validity. REGISTRY - identifies whether the structure of an
accession is unique to the Registry Structure file and updates
this file. NSC ASSIGNMENT - assigns an NSC Sample Number
depending on the uniqueness of the structure, supplier
information, and confidentiality. SSFID - identifies by iterative
search accessions which require special testing because they are
analogs of compounds of interest to the DTP Program. KEY
ASSIGNMENT - assigns structure fragments and chemically
descriptive keys to accessions to permit querying the file. CHEM
SEARCH INDEX - updates the search index with the assigned keys.
COMMUNICATIONS - generates action and informational reports
required by the accessioning and screening activities. BIO1,
BIO2, BIO3 and BIO4 - generate in a machine-processable form
varying data required by the Biological Data Processing System.
INVENTORY - enables the updating of inventory data and generates
manifests to ship new routine accessions for screening. CROSS
REFERENCE UPDATE - updates the Cross Reference file. The INQUIRY
and LOGX processors will be described as part of the Inquiry

Subsystem which follows.

Inquiry System

The CIS Inquiry subsystem was designed to be totally integrated into the processing system and to use state of the art techniques to allow users to regain the "feel" for the file lost in the previous batch system while retaining the flexibility and accuracy that iterative (atom by atom) searching of a detailed connection table allowed. The system developed uses data extracted from the structural and non-structural data files and indexes it in such a way that on-line, interactive searching from a remote terminal is practical.

Development of Search Keys. An analysis was made of previous DTP searches (16), and keys were developed that would make effective use of the detailed information content of the CAS Registry III connection table. Although some thought was given to developing novel substructure search keys, the reality of our development schedule led to the use of a combination of previously developed key types that have been found efficient. Figure 4 shows a mythical example of an abbreviated DTP structure record and sample of each type of search key. A description of each follows.

Augmented Atom (AA). This atom centered fragment is based on the concept used by CAS and others. (17, 18) Each non-hydrogen atom of a compound and its neighbors together with the corresponding bonding information generates an AA key and these large keys are exhaustively broken into smaller keys, the smallest being atom pairs. Occurrence counts are calculated and are associated with each NSC.

Ring System Keys. Since Registry III connection tables are segmented into an acyclic and a ring portion, ring keys are only assigned for the first occurrence of a ring system and stored. Thereafter, the appropriate ring keys are retrieved during key generation using the ring identifier. The following three keys are defined for every ring system in a compound:

Nucleus Keys (NUC). This key is generated from molecular formula of each complete cyclic nucleus in a substance. Two totals of certain bond types qualify this moleform: the sum of alternating and delocalized bonds; and the sum of tautomeric, double and triple bonds. Occurrence counts of a given key are also generated for each NSC and stored. In Figure 4, there is a total of six alternating bonds in the pyrimido ring of purine once these bonds have been normalized by Registry III algorithms. One double bond in the fused imidazo ring plus an alternating bond at the fusion point gives a (6,1) qualification to the element count.

ELEMENT KEYS (ELE)

ELE N   ELE O
ELE C   ELE S
      ELE H

RING KEYS (NUC, RIN, RSI)
(EXAMPLES)

TYPE	FORMULA	BONDS a/ A,D SUM	T,2,3 SUM
NUC	C5 N4	(6 ,	1)
RIN	C4 N2	(6 ,	0)
RIN	C3 N2	(1 ,	1)
NUC	C4 O	(0 ,	0)
RIN	C4 O	(0 ,	0)
RSI	5, 6		

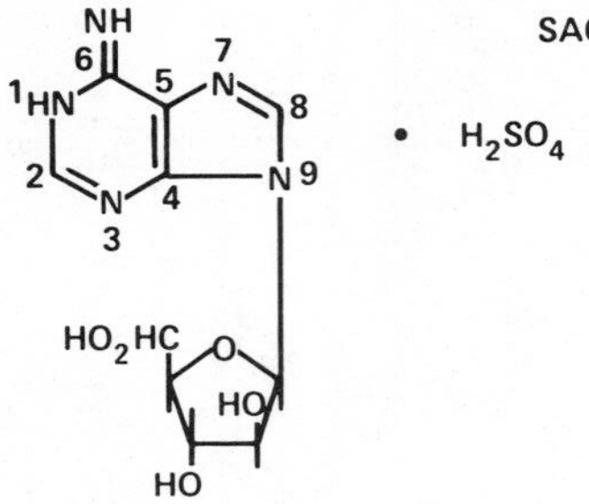

NSC-555555

*5 : β-D-ARABINO

SUPPLIER 240A-7

SAC

• H_2SO_4

EXTENDED MOLECULAR FORMULA (EMF)

$C10 \ H13 \ N5 \ O4 \ . \ H2 \ O4 \ S$

3 RINGS
6 DIRECT ATTACHMENTS TO RINGS
2 FRAGMENTS

AUGMENTED ATOM (AA)
(EXAMPLES AT ATOM 9)

	CENTRAL ATOM	AUGMENTING ATOM(S)	BONDS a/
AA	N	C C C	N1 R1 R1 (2)
AA	N	C C C	N1 R1
AA	N	C C C	R1 R1
AA	N	C	R1 (2)

a/ CAS BOND TYPES

A = ALTERNATING   1 = SINGLE
D = DELOCALIZED   2 = DOUBLE
C = TAUTOMER   3 = TRIPLE
           R = RING
           N = NON-RING

MISCELLANEOUS KEYS (MISC)

MISC D5	PRESENCE OF TYPE 5 CARBOHYDRATE OR AMINO ACID STEREODESCRIPTOR
MISC SUP 240A	SUPPLIER OF COMPOUND
MISC DD	DOT DISCONNECT
MISC TM	PRESENCE OF TAUTOMER
MISC SA	PRESENT ON SAC LIST OF 8000 AGENTS OF INTEREST

Figure 4. Examples of search keys

Ring Keys (RIN). This key cites the molecular formula for each individual ring in the smallest set of smallest rings (SSSR), plus those other rings of less than or equal to eight atoms. Bonding qualification is the same as in the NUC key.

RSIZE (RSI) Key. This key cites in ascending order the sequence of ring sizes in the SSSR. One RSIZE key per ring system is assigned although occurrence counts within a compound are kept. In Figure 4, the purine ring has an RSI=5,6 and furan has RSI=5.

Element Keys (ELE). One key is generated for the presence of each element in the periodic table plus deuterium and tritium. No occurrence counts are kept for this key.

Miscellaneous Keys (MISC). These are a mixture of non-structural and structural keys assigned from originally submitted data. As seen in Figure 4, miscellaneous keys are generated for such data as the presence of a type 5 stereodescriptor (carbohydrate or amino acid), a designation of DTP biological interest (SAC) and the supplier code of the submitter (sup 240A). In addition for compounds with no structure class designation, keys such as "Alkaloid" are assigned manually for compounds which cannot be machine registered.

Extended Molecular Formula (EMF). This search key consists of the molecular formula of each fragment in a compound, qualified by the number of prime rings in that moleform and the number of direct non-hydrogen atoms attached to rings.

Search File Organization. Search keys are stored in the Chemsearch Index File. To allow easy access to the approximately 15,000 keys generated by our open ended fragment generation and to give a user the opportunity to either use the extensive bonding information or ignore it, the keys are stored as inverted lists and are indexed using a variation of a hierarchical organization developed by Feldman (12). The very long lists are stored as bit maps for efficient intersection during Inquiry operation and shorter lists are stored in a compressed format which is converted to a bit map for intersection.

Searching the File. The search programs are divided functionally into two sections: definition and execution. The general philosophy is that all queries are defined on-line through interactive programs specifying search keys, optional iterative qualifications, and output statements and then stored. These questions can then be executed either on-line or in batch mode, with outputs being directed to the terminal, to system printers on the mainframe, or to two types of disk files (RESULTS and SAVE files) which can be accessed further by Inquiry. Both alphanumeric or structural data or a combination of each may be requested for output.

There are three main modes of Inquiry, each allowing different functions. Their system name and function follows:

GET. This mode allows searches by NSC Number, ranges of NSC Numbers, or by supplier code. The NSC Numbers can be input from a terminal or called from a prestored SAVE file.

Full Structure Search (FSS). This mode allows searches by molecular formula with optional qualification by number of rings, direct attachments to ring, number of dot disconnected fragments, or iterative search, i.e., atom–by–atom and bond–by–bond.

Substructure Search, (SSS). This mode allows searches by any combination of chemical or non-chemical keys or by a SAVE file used as a key. Output can be qualified by iterative search. A key logic statement is required using some combination of the keys already described.

Query Definition. The definition of a question is a two step procedure. Once the question has been defined by the user in his own mind in terms of information available in the system, component parts of the query such as the keys, any individual structures needed for iterative search, and a list of output data items are input interactively from a terminal, with edit programs validating each command. Once a component is input and stored under a user name in a file called the INQDFTN file, the components are assembled, using whatever logical relationship that is required, into a full query which is itself stored by name. This query can then be searched on–line using the terminal or run later, using a batch version of Inquiry.

An example of the prestored component pieces of an SSS query for Nucleosides and the resulting query is shown in Figure 5. The individual keys, structure and Data Item list and their identifying names are shown as well as an example of the later prompting in which the logical relationship of the components is stored as a question.

Structure S1 was entered using a variation of the query prompting language originally developed by Feldman (12) but modified and rewritten in PL-1. Short commands build a connection table in core with the corresponding picture of the numbered topological framework appearing on the screen. The user can then use terse commands corresponding to the numbered picture to indicate the level of specificity of each atom, using a variety of specific and generic atom types to qualify these by data (including hydrogen count, allowable unshown bonding, etc.) and to set a bond to any combination of the CAS bond types. The system defaults to certain bond types on the assumption that it is

STRUCTURE

S1:

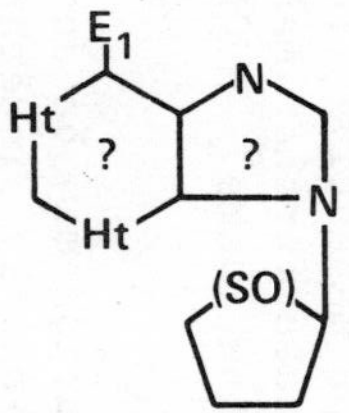

E₁ = N, O, S

PRESTORED SEARCH COMPONENTS
KEYS

K1: NUC C5 N(2,) MIN
K2: RIN C3 N2
K3: RSI 5,6 MIN
K4: RIN C4 S BOND A(O) N(O)
K5: RIN C4 O BOND A(O) N(O)

DATA

SDATA: FSUP, MCCO

(FIRST SUPPLIER AND
ACTIVITY INDICATOR)

ORGANIZATION OF COMPONENTS IN SEARCH

ENTER COMMAND: SSS NUCLEOSIDE
KEY LOGIC: S1 K1 K2 K3 (K4 OR K5)
ITERATIVE SEARCH (Y/N): Y
STRUCTURE LOGIC: S1
DATA ITEM LIST: SDATA
CHEM CARDS? (Y/N): N
SAVE FILE? (Y/N/#): Y
FILE SAVE 8 ASSIGNED
ENTER TERMINATION: END
SSS COMMAND COMPLETED

Figure 5. Formulation and storage of a substructure search question preferable to retrieve false responses than to lose valid ones, although these defaults can be overridden. Thus if a user specifies a ring double bond, the system will also allow tautomeric, alternating and delocalized bonds. Since the bonds in the example are set as "ring don't care", any ring bonds will be retrieved, including ring single.

The keys named K1 through K5 shown in Figure 5 are all designed to retrieve ring keys appropriate to the substructure desired. Ring keys are not assigned automatically from an iterative structure but instead are entered in a shorthand notation that allows user control over the degree of specificity in the key. Thus, key K1 will access any NUC Keys with five carbons (C5), two or more Nitrogens (N(2,)) plus any other elements (MIN), with any type of bonding allowed. K3 will retrieve any 5,6 ring regardless of fusion. While K4 and K5 specify exact ring formulae plus require that there be no unsaturated bonds (BON A(0) N(0)).

The Data Item list SDATA shown in Figure 5 specifies that data types FSUP (original supplier code), and MCCO (biology activity indicator) should be retrieved and printed with every answer.

Organization of the Query. Once the components are stored, a user invokes the formulation prompter for SSS and puts together logical relationship of components using the names stored in INQDFTN. The program interactively checks the logic for syntax, and determines if the components are actually in the INQDFTN file and are appropriate for the particular type of search being set up. As can be seen in Figure 5, the keys are put into a boolean expression ("AND" logic is the default) and the structure S1 is included to allow generation of augmented atoms at execution time. S1 is also the atom by atom expression that will be used to qualify the key hits.

The data item lists DATA specifies the two alphanumeric data items that will be retrieved and printed if present on the file. A SAVE file was specified here to permanently store the NSC Numbers or answers to the query in case further manipulation is necessary. The query is stored under the name "NUCLEOSIDE" for later access.

Inquiry Execution. Search execution may be done completely interactively, completely in batch, or partly interactively with completion in batch. All modes of search have a required key search in which the appropriate search index is accessed, the requested NSC numbers are retrieved, followed by an optional iterative search. In GET, this key search involves a simple lookup of the desired NSC numbers. In FSS the hierarchical extended molecular formula (EMF) index is scanned for particular moleforms plus any ring, attachments, and number of fragments specified. In SSS the key expression is analyzed and expanded from the shorthand notation used, the appropriate inverted lists accessed, intersected and a resultant list formed for further qualification.

The interactive execution of the search shown in Figure 5 would be invoked by the command "SEARCH NUCLEOSIDE". The structure S1 would generate AA keys and all possible miscellaneous and element keys. All permutations of indefinite elements such as "E1" and "SO" would be generated (although "Ht" for all hetero atoms, is too indefinite to do this). These augmented atoms (with any bond alternatives included in each key), the ring oriented keys, and any other keys are processed through the Chemsearch Index to retrieve individual specific inverted lists or bitmaps. These lists are then intersected according to the logical expression given. The program displays the number of key answers. The user has the option of cancelling the search or continuing with either fragment or iterative output. The output can be inspected on the terminal at a user-specified rate of "n" at a time. The user can end the search after this inspection or cancel print and route all output to a disk file.

Output. Based on what is requested in Query Definition, and on what a user commands during Inquiry Execution, the output of a search can be directed to an alphanumeric terminal, to a Tektronix graphics terminal if structures are needed, to CAS system printers, to a transaction file (PTFC) for the Communications Processor to produce standard chemistry cards at CAS, or to two types of disk files: a temporary browsing file (RESULTS); or a permanent SAVE file which can be accessed either sequentially through GET or as a search key in SSS.

Other Features. Although the main modes of Inquiry are GET, FSS, and SSS, another function of Inquiry called LOGX allows authorized Inquiry users to remotely access the last NSC number assigned by the CIS in its batch processing of new compounds. One or more new NSC Numbers or new Sample Numbers for previously-assigned NSC Numbers can be reserved to allow priority numbering of samples for compounds of interest so that they can be shipped for testing before registration in the CIS. The system also has a range of housekeeping commands to show, delete or modify stored items in INQDFTN to give aid to a user via "Help" files and to provide security via passwords and other protective measures.

Hardware and Software

The DR&D CIS is run at CAS on an IBM 370/168 under a virtual operating system (MVS). Other hardware used at CAS to support NCI structure input and output needs include: chemical typewriters, a CAS-developed refreshed graphics structure input system based on a PDP-15 and used in the On-Line Structure Input System (OLSIS) (15)

which can be used to show stereochemistry in structures; and a
non-impact off-line Varian/Statos printer used to produce all
large scale structure output requirements of the system resulting
from batch processing. As for all other NCI-developed programs in
the CIS, the Inquiry subsystem is written in PL-1 and uses
approximately 1200K bytes of main memory when run without
overlays. This memory requirement can be considerably reduced
when overlaying is used; however, Inquiry operates best without
the use of this segmentation. The telecommunications link is
accomplished through the IBM TCAM interface and currently three
ports are used to support terminals at NCI in Silver Spring, Md.,
at the storage and distribution contractor in Bethesda, Md., and
at CAS in Columbus, Ohio. At NCI the equipment used include a
Hazeltine 2000 CRT with 120 cps printer and tape cassette unit to
handle alphanumeric data and a Tektronix 4012 with hard copy unit
for vector graphics structure output. Both of these terminals can
be switched through a Bell Dataphone 1200 modem operating on a
leased foreign exchange line to Columbus at 1200 baud.

Interaction of the System with Other Data Bases

        While the DRD CIS was implemented primarily to process
compounds acquired for antitumor screening and to permit querying
the files on-line, the data base has been used for a variety of
drug development needs such as these described below.

        Name Index. Since the DR&D CIS uses the CAS Registry
techniques, the format of the DTP and CAS Registry Structure files
are identical. (To preserve the confidentiality of the NCI
compounds, however, the two files are separate and different check
digit algorithms are employed.) Building on this characteristic of
identical file formats, the Overlap Detection system was developed
to identify substances common to both files and to maintain a
cross reference of their CAS Registry Numbers to the DTP IRN and
thereby to NSC numbers. On a regular schedule, "non-systematic
names" in the CAS master name file are retrieved by algorithmic
examination and the CAS Registry numbers of these names passed
against the Overlap File to retrieve any corresponding NSC
numbers. These name records are then merged with a separate NCI
nonsystematic name file, sorted by name, and photocomposed to
print the Drug Development (DTP) Name Index. This index includes
all CAS and NCI "non-systematic names" and molecular formulas with
the corresponding CAS Registry number and NSC number, whichever of
these numbers is available. Thus, given a nonsystematic name, a
user of the index may be led to an NSC number and thereby to any
other data in the chemical or biological data files. If there is
no NSC, the CAS registry number can be used to look up the
systematic name in the CAS Registry Number Handbook or reference
other publications with Registry number indexes. The CAS Registry
Numbers are being added to the Cross Reference file of the DR&D

CIS to be retrieved as needed to facilitate access to the literature.

Literature Searching. Searches are made of the literature to identify compounds worthy of acquisition because of their reported biochemical or other biological properties or because of their structural characteristics. Chemical Abstracts Service's CASIA (and CA CONDENSATES) files were searched by profiles of DTP interest, e.g., compounds which inhibit DNA or RNA synthesis. The output is a listing containing the CA abstract number, indexing terms for the article, and for each compound in the article its CAS Registry Number, molecular formula, systematic name, and indexing terms as well as the corresponding NSC Number, if the substance was also registered in the DR&D CIS. (The NSC Number is obtained by computer match of CAS Registry numbers against the cross-reference file of CAS Registry/IRN maintained by the Overlap Detection procedures described above,) An inquiry of the DR&D CIS file for NSC Numbers so retrieved provides information such as the compound's antitumor activity or whether an additional supply is needed to complete testing. The structural diagrams for the compounds retrieved from the CA data base are generated by CAS' Algorithmic Structure Display (ASD) program from the connection table in the CAS Registry Structure file. After reviewing these data and the abstract (or original article, if needed), an acquisition effort is made for selected compounds. A similar but less automated approach is used when data bases other than CA are searched.

Substructure searches of CA were made with the cooperation of the Basel Information Center for Chemistry in Switzerland. The output is a listing of CAS Registry Numbers of compounds which meet the search criteria, their corresponding CA abstract number and a microfilm copy of the corresponding abstract. The structures of these retrieved compounds are input for a Search on Topology (SRT) against the DR&D CIS files to identify in that file compounds which are duplicates or near duplicates. The output of SRT, the abstracts (or articles of interest) and structure diagrams (generated by the automated Algorithmic Structure Display procedure) are reviewed by DTP to select compounds for acquisition. Substructure searches for acquisition purposes have also been made of other data bases such as the Index Chemicus Registry System maintained by the Institute of Scientific Information.

Preselection Subsystem. The following three "stand-alone" capabilities are available to aid medicinal chemists in making the optimal selection for acquisition from among available compounds: the Registry's Search on Topology (SRT) program which identifies duplicates and near duplicates; the Special Structure Feature Identification (SSFID) processor which identifies analogs of Program interest; and a statistical model (20) which calculates

the probability of antitumor activity of a compound based on an
analysis of its chemical fragments and antitumor activity in the
P388 lymphocytic leukemia screen. These three capabilities are
being improved and integrated into a preselection subsystem to
optimize acquisitions.

After processing potential acquisitions through SRT, search
keys would be assigned to the compounds and the data processed by
the SSFID processor which will be modified to increase its
efficiency. The SSFID processor will determine whether potential
acquisitions are in: chemical classes to be deemphasized because
they are largely negative or adequately studied; newly-emerging
active classes; active classes for which additional compounds are
desired; or classes hypothesized to yield active agents.

The probability-based model, as in other predictive methods,
uses a set of known active and inactive compounds as a training
set to derive a weight for activity and a weight for inactivity
for each structural feature fragment present from search key
assignment. Since the weight of each feature is based on
statistical significance, the activity score for a potential
acquisition is determined by summing the respective weights of the
structure features. A measure of rarity or uniqueness of
structural fragment present in candidate acquisitions is also
derived. The current model, being used with encouraging results
in an experimental mode, was based on a training set of 2,716
active compounds and 15,524 inactives. Work is ongoing to
evaluate the benefits of using additional or alternate chemical
features, such as Registry III Ring ID numbers or ganglia
augmented atoms (GAA) which include information on the bonds
extending from the terminal atoms.

Interlink between the Chemical and Biological Systems

The transfer of machine-processable data from the DR&D CIS to
BDPS regarding accessions, shipments, and replenishment actions
has been alluded to earlier. The data transferred includes:
names and addresses of suppliers, screeners, and the DTP staff
which is used by the BDPS to mail the summarized biological test
results; NSC Numbers assigned to new accessions and other
substance-related data such as chemical class which is used to
build the BDPS master control record to anticipate test data and
to report the reason for acquisition and special testing; and out
of stock and replenishment actions which are reported to DTP
decision makers. Further, NSC Numbers retrieved by Inquiry from
the DR&D CIS may be written onto a cassette unit which is part of
the Hazeltine terminal configuration and transformed into
transactions for querying the BDPS to retrieve the test results
for these substances. However, there is no tranfer of evaluated
test results or other data from the BDPS to the DR&D CIS.

It should be noted that the DR&D CIS data base is resident in
Columbus, Ohio and is queried in an on-line interactive mode,

while the BDPS is resident in Bethesda, Maryland in NIH's Division of Computer Research and Technology, and it's serial tape files queried in batch. There is as yet no capability of producing a report combining selected data from both data bases let alone browse in both data bases, regardless of the system in which the query originated. The fact that we have recently generated structures by ASD for the bulk of the 227,000 compounds registered before the start of the DR&D CIS is a significant step in interfacing these systems more feasible. However, new hardware, software and communications are required before the chemical and biological data bases may be interfaced with improved effectiveness.

Acknowledgements

Special mention is due to: David Lefkovitz, Helen Hill, Margaret Milne, and Ruth Powers of the University of Pennsylvania for their efforts in the design and implementation of the DR&D CIS; Warren S. Hoffman and his team of programmers of E.I. du Pont De Nemours and Company for their early programming support; and Gary Kurtenbach and Charles Watson of the Chemical Abstracts Service for their support in system installation and operation.

Abstract

This automated system was developed to monitor the accession, storage, and distribution of chemicals and drugs acquired for antitumor testing. The system identifies duplicates and analogs of Program interest, produces action and informational documents required by the drug development activity, maintains inventory control, interfaces with the Biology Data Processing System which evaluates and disseminates screening results, and tracks acquisitions from receipt through shipment to the screening laboratories and the return of the excess supply. An on-line, interactive retrieval subsystem permits substructure and full structure searching on a file of 310,000 compounds. Results of searching other data bases for structural moieties and biological properties are linked with the system to identify compounds worthy of acquisition. An index of nonsystematic names in the Chemical Abstracts referenced to compounds in the system is generated.

Literature Cited

1. Zubrod, C.G., Schepartz, S., Leiter, J., Endicott, K.M., Carrese, L.M., Baker, C.G., "The Chemotherapy Program of the National Cancer Institute: History, Analysis, and Plans," Can Chem. Rep., (1966), 50, 349-540.

2. Ihndris, R.W., "Chemical Structure Fragmentation for Use in a Coordinate Index Retrieval System," J. Chem. Doc., (1964), 4, 274-278.

3. Leiter, D.P., Morgan, H.L., Stobaugh, R.E., "Installation and Operation of a Registry for Chemical Compounds," J. Chem. Doc., (1965), 5, 238-242.

4. Morgan, H.L., "The Generation of a Unique Machine Description for Chemical Structures - A Technique Developed at Chemical Abstracts Service," J. Chem. Doc., (1965), 5, 103-113.

5. Hazard, G.F.Jr., Murray, B.R., "The Use of the Chemical Abstracts Registry System in a Drug Development Program," 7th Annual Middle Atlantic Regional Meeting of the American Chemical Society, Philadelphia, Pa., 1972.

6. Hazard, G.F.Jr., "The Use of the CAS Substructure Search System on a Large Drug Development Chemical File," presented at the 162nd Meeting of the American Chemical Society, September, 1971.

7. Wigington, R.L, "Machine Methods for Accessing Chemical Abstracts Service Information," 97-120, in "Proceedings of IBM Symposium on Computers and Chemistry," IBM DATA Processing Division, White Plains, New York, 1969.

8. First and Second Updates to the CAS Facility for Integrated Data Organization, Users Manual, 1974 NTIS PB 242-856

9. Ditmar, P.G., Stobaugh, R.E., Watson, C.E., "The Chemical Abstracts Service Chemical Registry System. I. General Design," J. Chem Info. Comput. Sci., (1976), 16, 111-121

10. Freeland, R.G., Funk, S.J., O'Korn, L.J., Wilson, G.A., "Augmented Connectivity Molform - A Technique for Recognition of Structure Topology Identity," presented at the 169th National Meeting of the American Chemical Society, Philadelphia, Pa., April, 1975.

11. Blackwood, J.E., Elliott, P.M., Stobaugh, R.E., Watson, C.E., "The Chemical Abstracts Service Chemical Registry System. III Stereo-chemistry," J. Chem. Info. Comput. Sci. (1977), 17,3-8

12. Feldmann, R.J., Heller, S.R., "An Application of Interactive Graphics-The Nested Retrieval of Chemical Structures," J. Chem. Doc., (1972), 12, 48, 48.

13. Powers, R.V., Hill, H.N., "Designing CIDS - The U.S. Army Chemical Information and Data System," J.Chem. Doc., (1971), 11, 30-38.

14. Dittmar, P.G., Mockus, J, Courvreur, K.M., "An Algorithmic Computer Graphics Program for Generating Chemical Structure Diagrams," J. Chem. Info. Comput. Sci., (1977), 17, 186-192.

15. Blake, J.E., Farmer, N.A., Haines, R.C., "An Interactive Computer Graphics System for Processing Chemical Structure Diagrams," J. Chem. Info. Comput. Sci., (1977), 17, 223-228.

16. Milne, M., Hazard, G.F.Jr., "A Systematic Analysis of Substructure Search Questions for the Chemotherapy Program of the National Cancer Institute, 164th National Meeting of the American Chemical Society, August, 1972.

17. "Substructure Search System Documentation," Chemical Abstracts Service, Columbus, Ohio, 1970.

18. Adamson, G.W., Lynch, M.F., Town, W.G., J. Chem. Soc.-C, 3702, 1971.

19. Milne, M., Hazard, G.F.Jr., "National Cancer Institute Drug Research and Development Chemical Information System: Substructure Search," 169th National Meeting of the American Chemical Society, April, 1975.

20. Hodes, L., Hazard, G.F.Jr., Geran, R.I., Richman, S., "A Statistical-Heuristic Method for Automated Selection of Drugs for Screening," J. Med. Chem., (1977), 20, 469-475.

RECEIVED August 29, 1978.

INDEX

DATE DUE

CARR McLEAN, TORONTO FORM #38-297